A Maths Dictionary

A Maths Dictionary

Second edition

For IGCSE

R.E. Jason Abdelnoor
Sue Chandler

Nelson Thornes

Introduction

Mathematics is a language. To think mathematically or express yourself in mathematical terms you must first learn the correct meaning of the words you need to use.

The function of this dictionary is to explain the mathematical terms appearing in maths books and projects, covering the secondary years up to the examination at 16+. It is a handy reference tool for the classroom, for the library and the home bookshelf.

How to use this book

The mathematical terms are set out alphabetically as in a normal dictionary. To understand one mathematical term you will often have to understand others with which it is linked.

Cross-references to other terms that appear as separate entries are italicised. (In certain instances the italicised word may not match precisely the actual word defined, e.g. ***graphical*** would be intended to indicate the entry **Graph**.) So in the definition for **Abacus**, ***base*** is defined separately in the dictionary.

To emphasise a particular feature in a diagram or an explanation, various words, lines or figures are highlighted in green.

A list of symbols with their definitions is given on pages vi–vii.

Mathematical symbols

SYMBOL	*DEFINITION*
Sets	
U	universal set
$\{\ \}$ or $\varnothing$	the null (empty) set
$\subset$	a subset of
A′	complement a set of A
{x: …}	the set of all x such that …
$\cup$	union of sets
Relations and functions and graphs	
$y \propto x^n$	y varies as x^n
$gf(x)$	$g[f(x)]$
$g^2(x)$	$g[g(x)]$
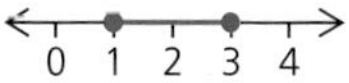	$\{x: 1 \leqslant x \leqslant 3\}$
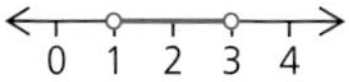	$\{x: 1 < x < 3\}$
Number Theory	
W	the set of whole numbers
$\mathbb{N}$	the set of natural (counting) numbers
$\mathbb{Z}$	the sets of integers $\begin{Bmatrix} \mathbb{Z}^+ \text{ – positive integers} \\ \mathbb{Z}^- \text{ – negative integers} \end{Bmatrix}$
$\mathbb{Q}$	the set of rational numbers
$\mathbb{R}$	the set of real numbers
$5.\dot{4}3\dot{2}$	5.432 432 432
$9.87\dot{2}\dot{1}$	9.872 121 212
Measurement	
05:00 h.	5:00 a.m.
13:15 h.	1:15 p.m.
7 mm ± 0.5 mm	7 mm to the nearest millimetre
10 m/s or 10 ms^{-1}	10 metres per second

SYMBOL	*DEFINITION*
Geometry	
M	reflection
R_θ	rotation through $\theta°$
T	translation
G	glide reflection
E	enlargement
MR_θ	rotation through θ followed by reflection
$\measuredangle$, $\angle$, $\wedge$	angle
$\equiv$	is congruent to

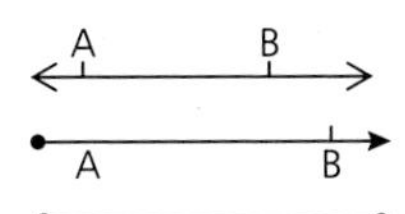

line AB

ray AB

line segment AB

Vectors and Matrices	
$\underline{a}$ or **a**	vector a
$\overrightarrow{AB}$	vector AB
$\lvert\overrightarrow{AB}\rvert$	magnitude of vector AB
If	$\begin{bmatrix} a & b \\ c & d \end{bmatrix}$ or $\begin{pmatrix} a & b \\ c & d \end{pmatrix}$ is the matrix X
then	$\begin{vmatrix} a & b \\ c & d \end{vmatrix}$ is the determinant of X, written $\lvert X \rvert$ or det x.
A^{-1}	inverse of the matrix A
I	identity matrix
O	zero matrix
Other Symbols	
$=$	is equal to or equals
$\geqslant$	is greater than or equal to
$\leqslant$	is less than or equal to
$\simeq$	is approximately equal to
$\Rightarrow$	implies
$A \Rightarrow B$	If A, then B
$A \Leftrightarrow B$	If A, then B and If B, then A } A is equivalent to B

To my wife, Jeannie, without whose loyal
support this book would not have been written

Abacus

An abacus is apparatus used for counting.
Examples: These abaci are all used for counting in our ***base*** ten number system.

Child's abacus

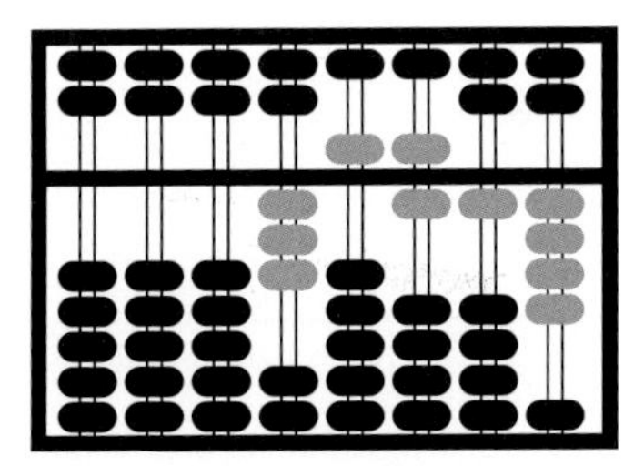

Chinese abacus showing the numbers 00035614

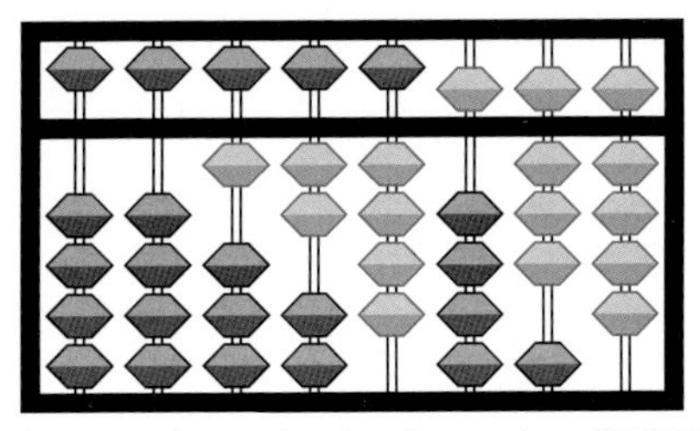

Japanese abacus showing the numbers 00124589

Absolute error

The absolute error is the ***difference*** between the real value of a quantity and an ***estimated*** value.

Example: When a number is estimated as 6 whose true value is 6.5, the absolute error is 0.5.
See **Error, Percentage error, Relative error**

Acceleration

The acceleration of a body is the change in ***speed*** per ***unit*** of time.

Examples: A car is accelerating uniformly along a straight road and its speed is recorded every 5 seconds, as shown in this table:

speed, m/s	6	12	18	24	30
time, s	0	5	10	15	20

The constant acceleration over this period is

$$\frac{6\text{ m/s}}{5\text{s}} = 1.2\text{ m/s}^2$$

A sprinter in the Olympic Games reaches a speed of 36 km/h in the first 4 seconds of his race. His average acceleration over the first 4 seconds is

$$\frac{36\text{ km/h}}{4\text{ s}} = \frac{36000\text{ m/h}}{4\text{ s}}$$

$$= \frac{36000\text{ m/s}}{60 \times 60 \times 4\text{ s}}$$

$$= 2.5\text{ m/s}^2$$

Acute

Acute means sharp. An acute ***angle*** is a sharply pointed angle whose size is between 0° and 90°.

Examples:

Additive inverse

The additive inverse is the ***inverse*** under addition. It is the number which, when added to the given number, gives the ***identity element*** for addition.

Examples:
The additive inverse of 3 is -3, because $3 + (-3) = 0$
The additive inverse of 5 is -5, because $5 + (-5) = 0$
The additive inverse of -6 is 6, because $-6 + 6 = 0$
The additive inverse of $-\frac{1}{2}$ is $\frac{1}{2}$, because $-\frac{1}{2} + \frac{1}{2} = 0$

NOTE: 0 is the identity for addition of numbers.

With ***vectors*** and ***matrices*** similar rules apply.

The additive inverse of $\begin{pmatrix} 3 \\ -5 \end{pmatrix}$ is $\begin{pmatrix} -3 \\ 5 \end{pmatrix}$, because $\begin{pmatrix} 3 \\ -5 \end{pmatrix} + \begin{pmatrix} -3 \\ 5 \end{pmatrix} = \begin{pmatrix} 0 \\ 0 \end{pmatrix}$

The additive inverse of $\begin{pmatrix} 1 & 0 \\ -2 & 4 \end{pmatrix}$ is $\begin{pmatrix} -1 & 0 \\ 2 & -4 \end{pmatrix}$,
because $\begin{pmatrix} 1 & 0 \\ -2 & 4 \end{pmatrix} + \begin{pmatrix} -1 & 0 \\ 2 & -4 \end{pmatrix} = \begin{pmatrix} 0 & 0 \\ 0 & 0 \end{pmatrix}$

Altitude of a triangle

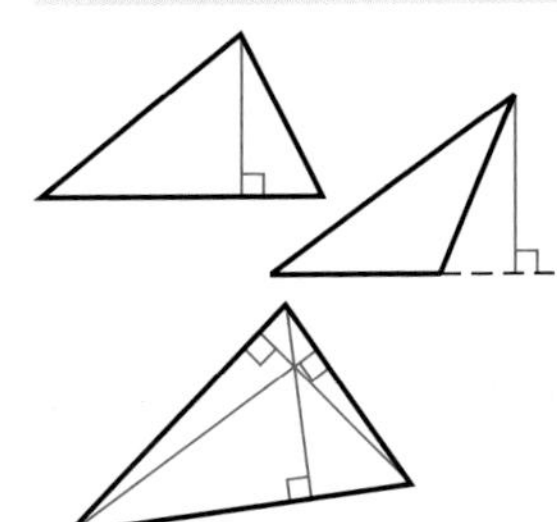

An altitude is the ***line*** from a ***vertex*** of a ***triangle*** drawn ***perpendicular*** to the side opposite the vertex.

Examples (left): Some altitudes are shown in green.
An altitude may be inside or outside the triangle. There are three altitudes for each triangle, one from each vertex, and they all pass through the same point.
See **Orthocentre**

Amount

The amount is the ***sum*** of the ***principal*** and ***interest***. The interest is usually ***compound interest*** but it can be ***simple interest***.

Examples: If £1000 is invested for five years and with the added interest becomes £1450 then

£1450 = £1000 + £450
amount = principal + interest

Amplitude

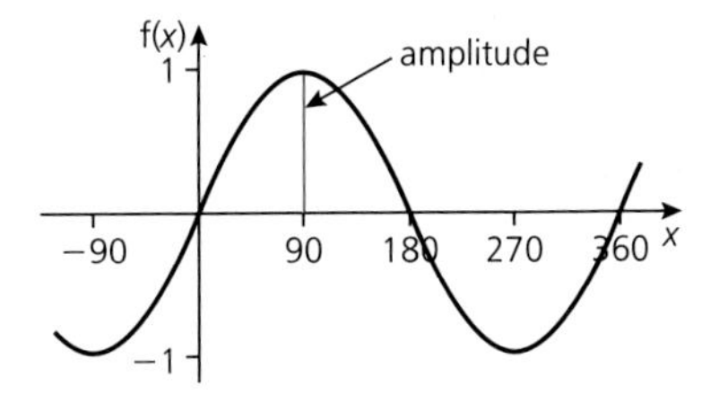

The amplitude of a ***periodic*** curve is half the value of the difference between the ***maximum*** and ***minimum values*** of the ***function***.

Example: The amplitude of $f(x) = \sin x$ is 1.

Angle

Angle between lines

An angle is the amount of turn from one ***line*** to another.

Examples: When any two straight lines meet at a point they form an angle. The point where the lines meet is called a ***vertex***.

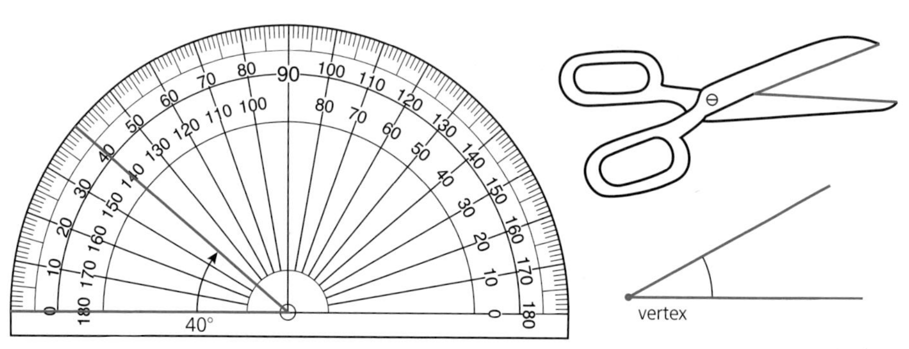

Angles are measured in degrees using a protractor.

Angles of a polygon

The angles of a ***polygon*** are the angles between the sides.

Examples: In the figure ABCD we can refer to angle x as $\hat{A}$ or $\angle A$; but for angle y we must refer to it as $A\hat{B}D$ or $\angle ABD$ not just $\hat{B}$ or $\angle B$ to avoid ambiguity.

The angles of the two polygons below are shown in green.

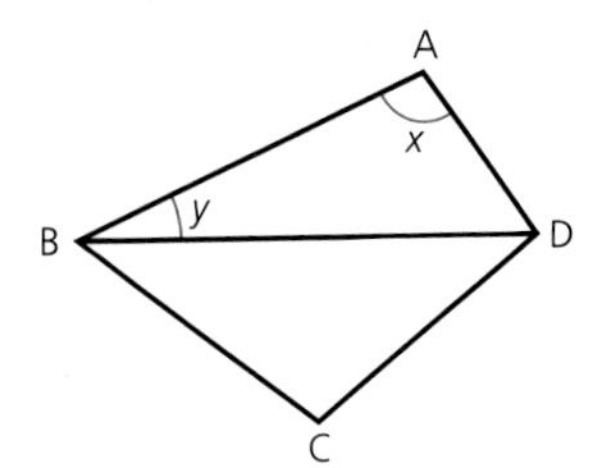

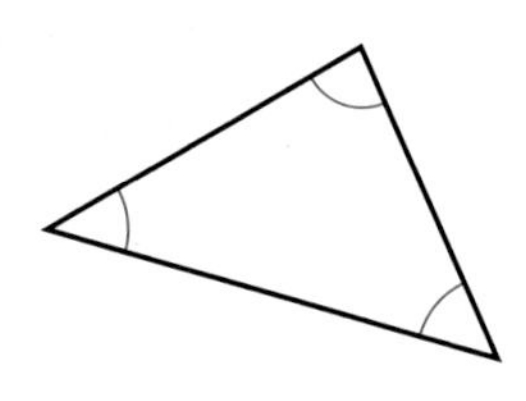

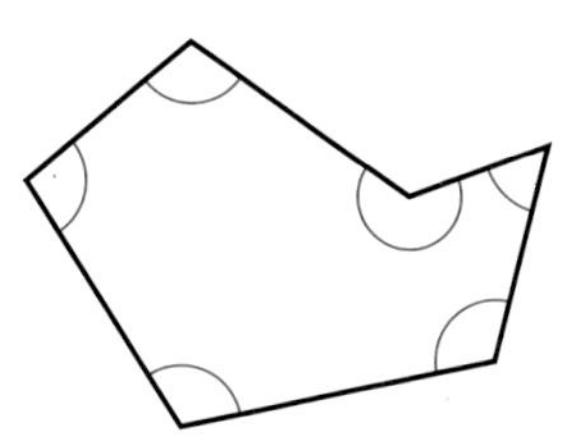

Direction of angles

Clockwise: this is the direction in which the hands on a clock turn.

Anticlockwise: this is the opposite direction.

Bearings are measured clockwise

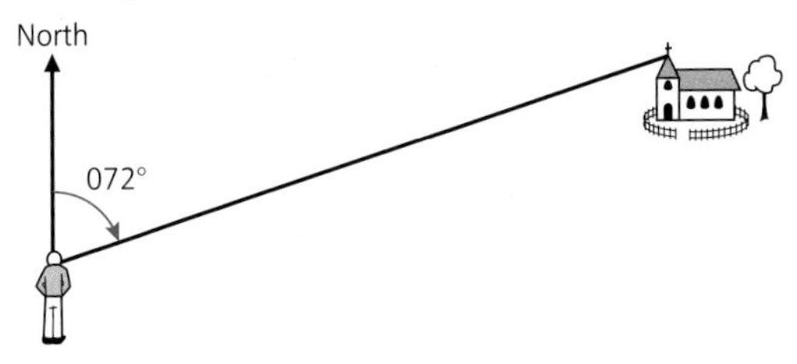

Rotations are called ***positive*** when they are anticlockwise, and negative when they are clockwise.

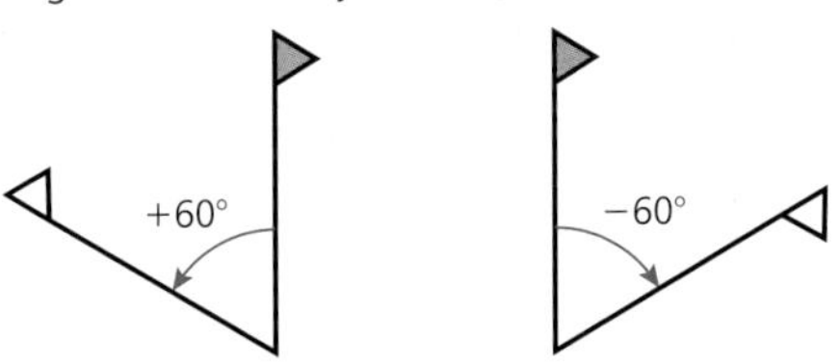

Angle bisector see **Bisector**

Anticlockwise see **Angle**

Apex see **Pyramid**

Approximation

A value that is near but not equal to the real value.

Example: $\frac{22}{7}$ is an approximation to the real value of π.

The word is similar in meaning to ***estimation***.

Arc

An arc is a ***line***, straight or curved, which joins two points.

NOTE: In a ***circle***, a part of the ***circumference*** is called an arc. The green part is an arc of the circle.

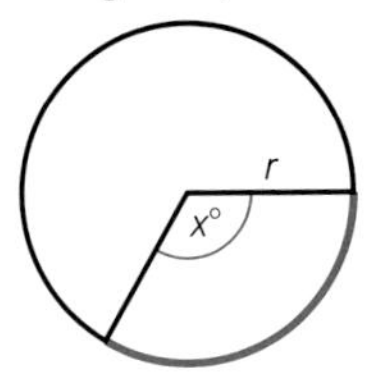

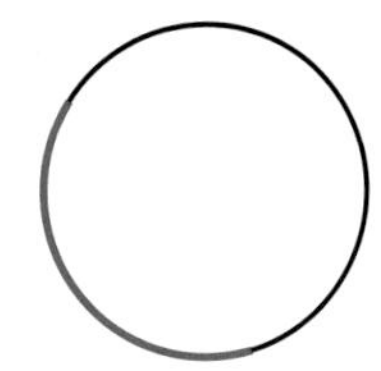

The length of the arc $= \frac{x}{360} \times 2\pi r$

Area

The area is the size of a surface. The surface may be ***plane*** (flat) or curved.

Here are some ways of finding areas of important shapes:

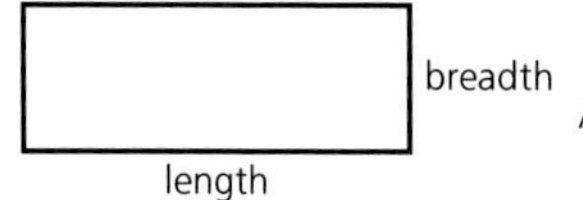

Area of ***rectangle*** = Length × Breadth.

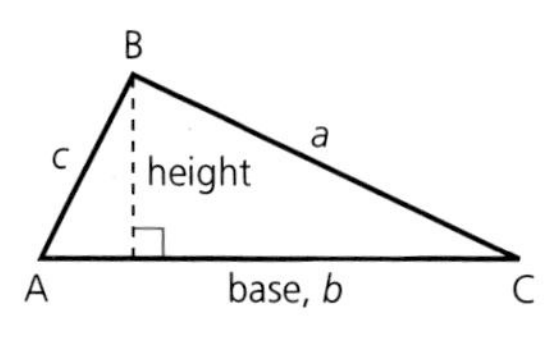

Area of ***triangle*** $= \frac{1}{2}$ Base × Height.

$= \frac{1}{2}ab \sin C$

$= \sqrt{s(s-a)(s-b)(s-c)}$ where $s = \frac{1}{2}$ the ***perimeter***

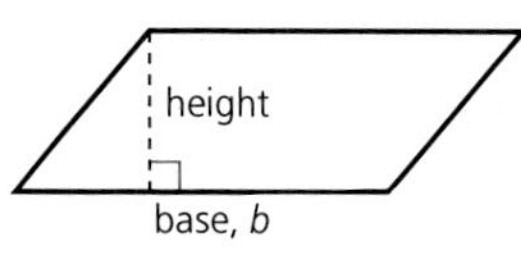

Area of ***parallelogram*** = Base × Height.

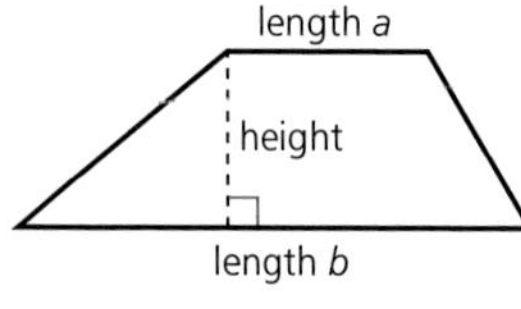

Area of ***trapezium*** $= \frac{1}{2} \times (a + b) \times$ Height.

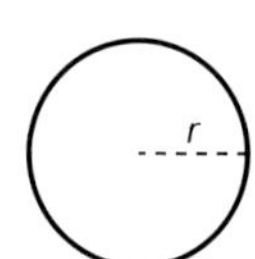

Area of a ***circle*** $= \pi r^2$.

Arithmetic mean

The arithmetic mean is what most people call the 'average'.
See **Mean**.

Arithmetic progression

An arithmetic progression is a ***sequence*** of numbers where the ***difference*** between successive ***terms*** is ***constant***. The difference between the successive terms is called the common difference.

When the first term is a and the common difference is d, the sequence is

$a, (a + d), (a + 2d), (a + 3d), \ldots$

The sum of the first n terms is $\frac{1}{2}n(2a + (n - 1)d)$

Arrow diagram

An arrow diagram shows a ***relation*** between ***sets*** of things, such as people or numbers. See **Correspondence**

Examples:

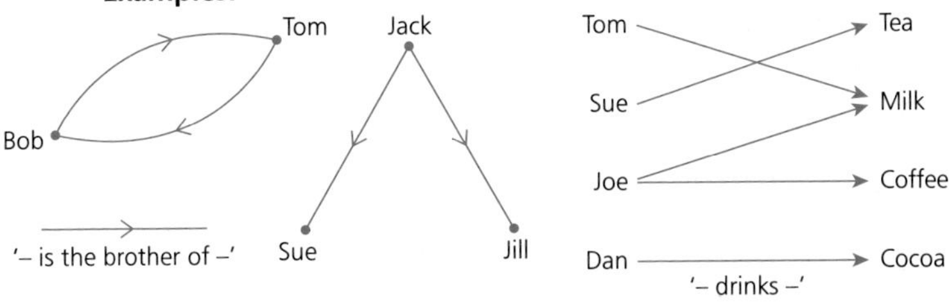

Associative

An ***operation*** is said to be associative if it does not matter where ***brackets*** are put when three ***elements*** are combined.
$(a*b)*c = a*(b*c)$ is true if $*$ is associative.

Examples:

Numbers under {addition and multiplication} ARE associative

$(2 + 3) + 5 = 2 + (3 + 5)$
$(4 \times 2) \times 3 = 4 \times (2 \times 3)$

Numbers under {***subtraction*** and ***division***} are NOT associative

$(4 - 3) - 1 \neq 4 - (3 - 1)$
$(12 \div 6) \div 2 \neq 12 \div (6 \div 2)$

Matrices under {addition multiplication} ARE associative.

Average

A teacher might ask for a typical member of the class to represent the group at chess, or football. An average is one member, or value, that represents the whole group. Mathematicians calculate three very important averages; they are ***mean***, ***median***, and ***mode***.

Axis (plural **Axes**)

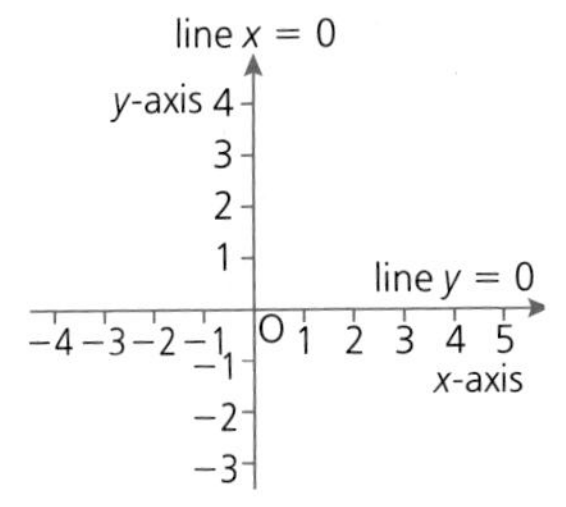

Axes are the reference ***lines*** on a ***graph***. On a ***coordinate*** graph, the line going across has the x numbers and so is called the x-axis. The line going up the page has the y numbers and so is called the y-axis. The point where the lines cross is called the ***origin*** and represents ***zero*** on both axes.

NOTE: The ***equation*** of the x-axis is $y = 0$ because the y-coordinate is 0 for every point on it. Similarly the y-axis has equation $x = 0$.
See **Cartesian plane**

B

Bar chart

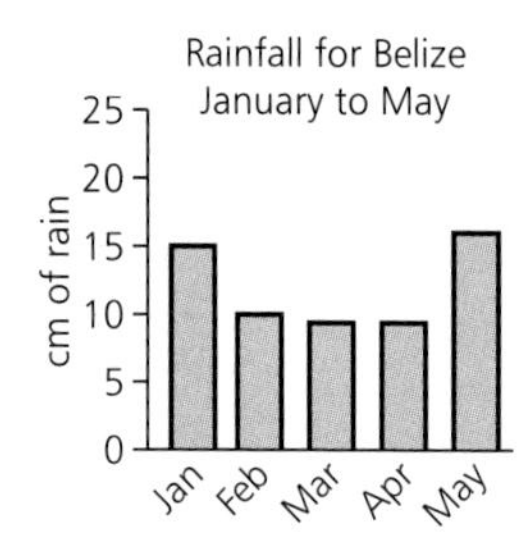

Bar charts show information in a ***graphical*** form by the use of columns or bars. The length or height of the columns or bars corresponds to the size they represent. Rainfall graphs are often bar charts.

Base

When counting things we use groups.

Example (left): There are two groups of 5 sheep and 3 sheep on their own. So the number of sheep is here represented as 23_{five}. The base of a number is the size of the group used. The diagram shows counting in base five. The number of sheep is 23_{five}.

Examples: Our normal counting number system is called denary or base ten because the group size used is ten.

We use ten ***digits***, 0, 1, 2, 3, 4, 5, 6, 7, 8, 9.

The place values are ***multiples*** of ten so that

the 6 in 7614 stands for 6×100
the 6 in 8563 stands for 6×10
352 stands for $(3 \times 100) + (5 \times 10) + (2 \times 1)$

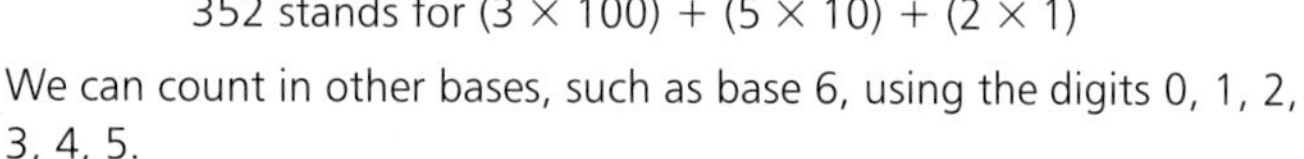

We can count in other bases, such as base 6, using the digits 0, 1, 2, 3, 4, 5.

The group sizes are 6, 6×6 (= 36), $6 \times 6 \times 6$ (= 216)
so that 3425 stands for $(3 \times 216) + (4 \times 36) + (2 \times 6) + (5 \times 1) =$
$648 + 144 + 12 + 5$ (in base ten) $= 809_{ten}$

Bearing

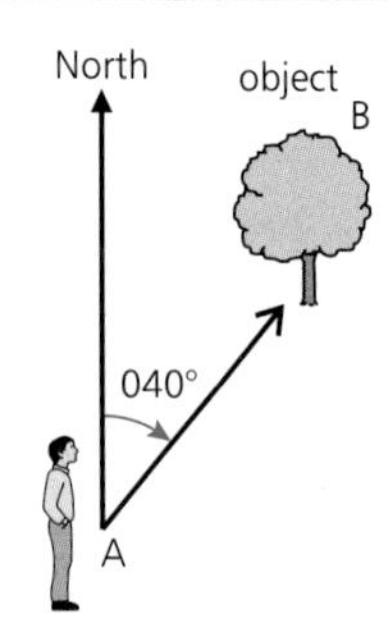

The three-figure bearing of an object is the ***angle*** measured ***clockwise*** from north to the object.

Bearings always have three ***digits***: 040° NOT JUST 40°.

Examples:

Bearing of B from A is 040°

Bearing of C from D is 230°

object C

North

D

230°

See **Compass direction**

Bilateral symmetry

A shape has bilateral symmetry when it has one ***line of symmetry***. When an object is reflected in a ***mirror line***, the ***object*** and its ***image*** make a shape that has bilateral symmetry.

Binary numbers

Binary means two, so binary numbers are numbers in ***base*** two. In the binary system only two ***digits*** are used: 0 and 1. The ***place values*** are

..	..	..	32	16	8	4	2	1
						1	0	$1_{two} = 5_{ten}$
					1	1	0	$0_{two} = 12_{ten}$
				1	1	0	1	$1_{two} = 27_{ten}$

Binary operation see **Operation**

Bisector

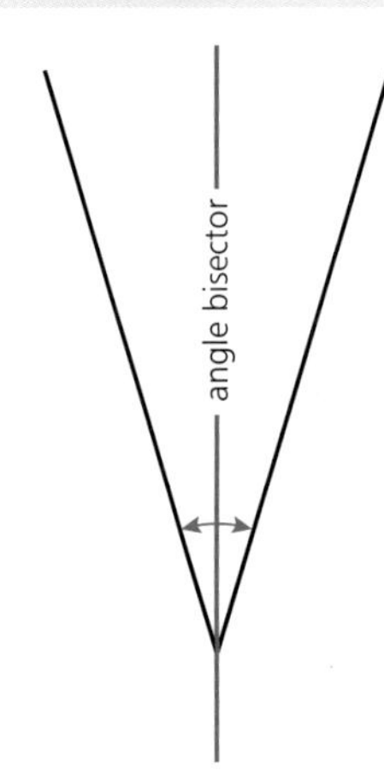

A bisector cuts something into two equal parts.

Examples:

a) A ***line*** which cuts an ***angle*** into two equal parts is called an angle bisector.

b) One line is a bisector of another line if it cuts it into two equal parts.

Line p is a bisector of line AB.

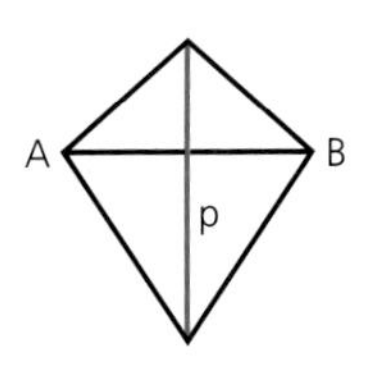

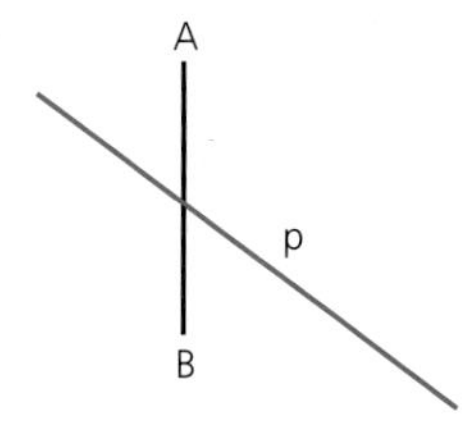

If the bisector of AB is also at ***right-angles*** to AB, the bisector is called the ***perpendicular bisector*** of AB. This line is sometimes called the ***mediator*** of AB.

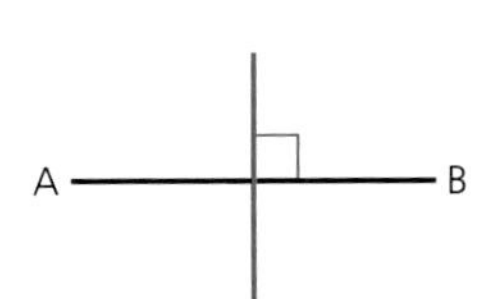

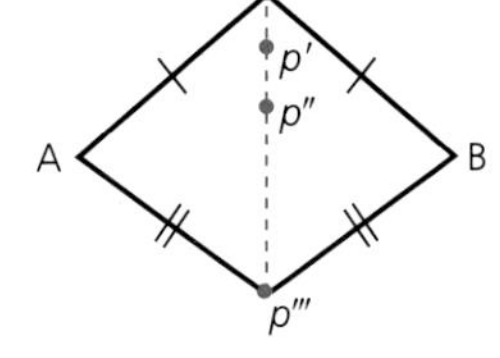

NOTE: The mediator of AB is the ***locus*** of points p which are equidistant from A and B.

Boundary

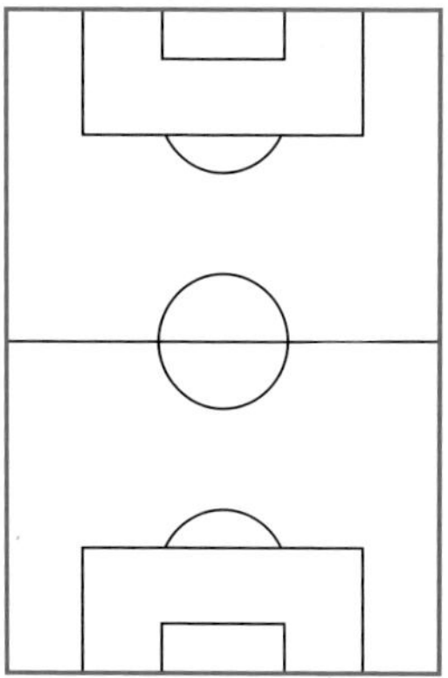

football pitch

The boundary of a figure is the ***line*** around the outside of the figure. The length of the boundary is called the ***perimeter***.

Examples: The boundaries are shown in green.

outline map of USA

Brackets

Brackets are used when three or more ***elements*** are combined by ***operations***. Brackets tell you which two elements you combine first.

Examples: $(10 - 7) + 2 = 3 + 2$
$= 5$

but

$10 - (7 + 2) = 10 - 9$
$= 1$

C

Cancel

Cancelling a ***fraction*** is finding an ***equivalent fraction*** with smaller numbers. See **Equivalent fraction**

Examples:

$\frac{12}{15} = \frac{\not{3} \times 4}{\not{3} \times 5}$ which cancels to $\frac{4}{5}$

$\frac{14}{21} = \frac{\not{7} \times 2}{\not{7} \times 3}$ which cancels to $\frac{2}{3}$

$\frac{24}{32} = \frac{\not{8} \times 3}{\not{8} \times 4}$ which cancels to $\frac{3}{4}$

Capacity

Capacity is a measure of ***volume*** used to describe how much a container (a can, for example) can hold, usually of liquid.
A ***litre*** is a ***unit*** of capacity, and 1 litre = 1000 ***cubic centimetres***.

Cardinal number

Cardinal numbers describe the number of objects in a collection. The number of ***elements*** in a ***set*** is a cardinal number. The cardinal numbers are 0, 1, 2, 3, ...

Cartesian plane

The Cartesian plane is a ***plane*** where the position of a point on it is given by its ***displacements*** from two straight ***lines***, called ***axes***. These axes are usually ***perpendicular***.

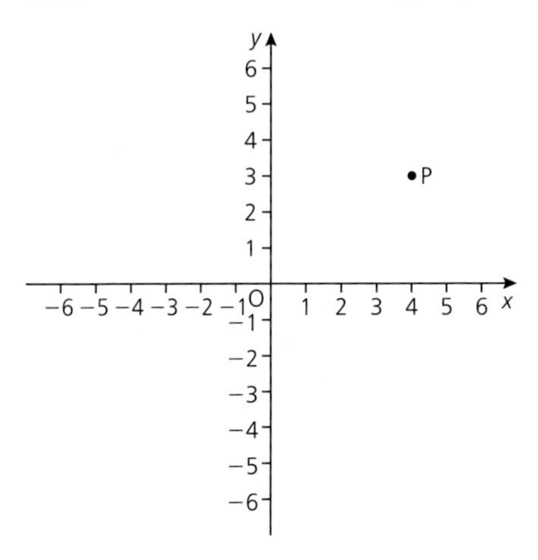

The position of point P in the Cartesian plane is given by the ***coordinates*** (4, 3).

Centimetre

A centimetre is one-hundredth of a ***metre***.
1 cm = $\frac{1}{100}$ m and is this long (1 cm)
NOTE: We shorten centimetre to cm.

Examples:

This matchstick is 4 cm long.

4 cm

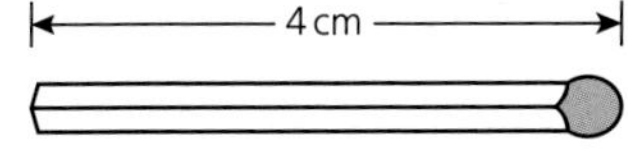

This ***line*** is 5 cm in length.

Central tendency

A measure of central tendency is either the ***mean***, the ***median*** or the ***mode*** of a set of numbers.

Centre of rotation

The centre of rotation is the one point which does not move when you do a ***rotation***. Where tracing paper is used the centre of rotation is the point where you put your pen or compass point.

Examples:

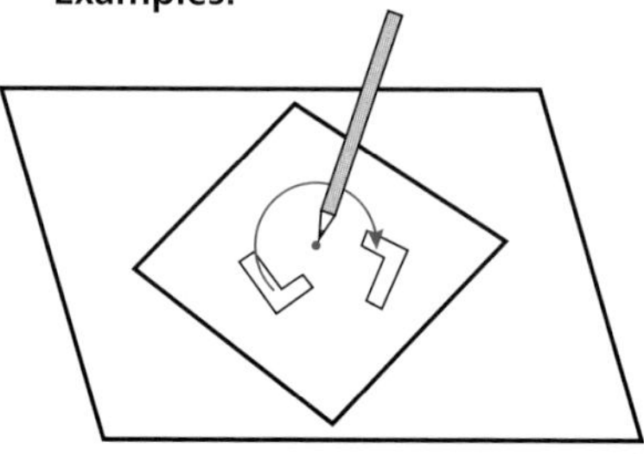

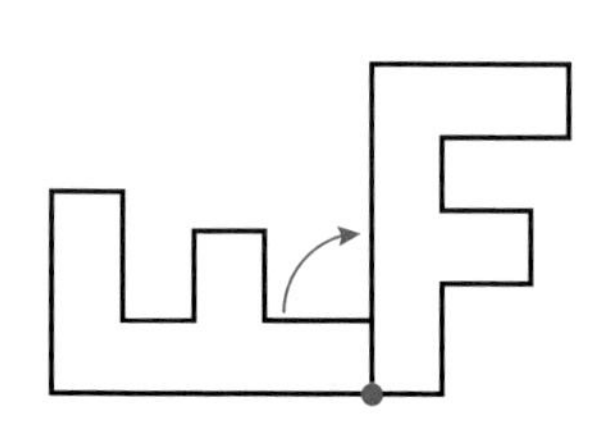

The centres of rotation are shown in green.

Centroid

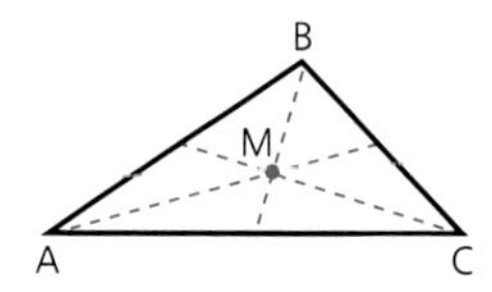

The centroid of a ***triangle*** is the point where the ***medians of the triangle*** intersect.

Example: The point M is the centroid of the triangle ABC.
See and compare with **Orthocentre**.

Characteristic

The characteristic is the ***whole number*** part of a ***logarithm***.

Example: log 400 = 2.6021; 2 is the characteristic and 0.6021 is the ***mantissa***.

Chord

A ***line*** joining two points of a ***circle*** is called a chord.

Examples:

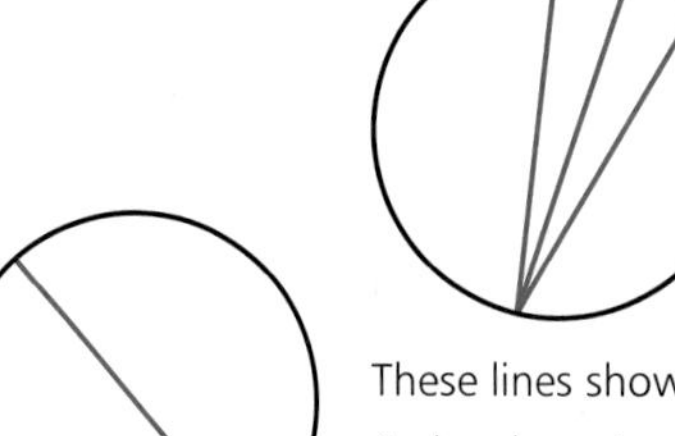

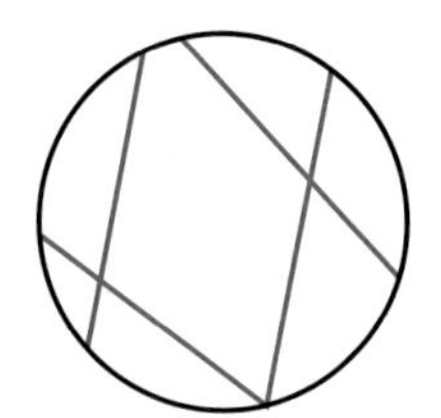

These lines show just SOME of the chords of these circles.

A chord passing through the centre of the circle is called a ***diameter***.

There is only ONE diameter through A.

Circle

A circle is the ***set*** of all points in a ***plane*** at a fixed distance, the ***radius***, from a fixed point, the centre.

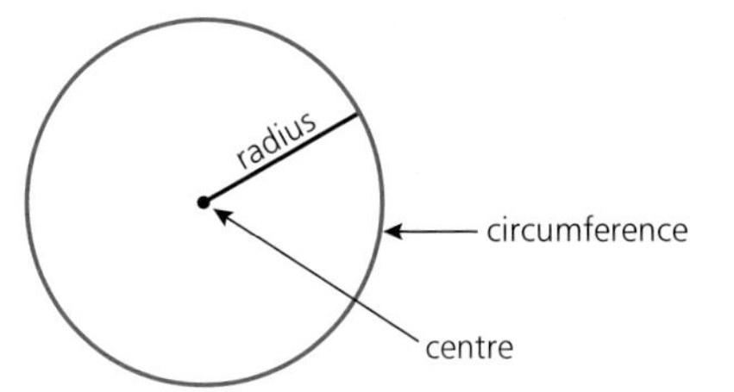

Example (left): The ***locus*** of a hammer being swung on the end of a chain is a circle.

For the ***area*** inside a circle, see **Pi**.

See also **Arc**, **Circumference**, **Sector**, **Segment**, **Semicircle**

Circumference

The circumference of a ***circle*** is the distance right around the circle. It is a special name for the ***perimeter*** of a circle.

$$\text{Circumference} = \pi \times \textbf{\textit{diameter}}$$
$$C = \pi d \text{ (exactly)}$$

It is roughly 3 times the diameter.

$$C = 3 \times d \text{ (approximately)}$$

For π see **Pi**

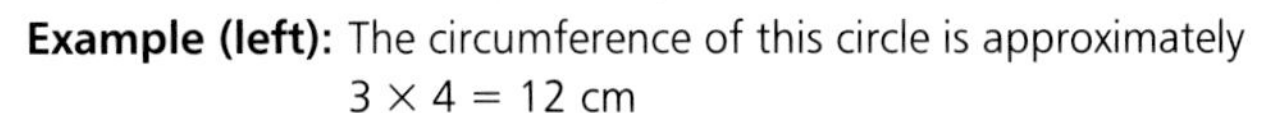

Example (left): The circumference of this circle is approximately
$$3 \times 4 = 12 \text{ cm}$$

The circumference of a circle is also the distance rolled by a circle before coming the same way up again.

The circumference of this circle is shown by the length of the green ***line***.

Circumscribe

One geometric shape circumscribes another when it is drawn round and touches but does not cross the second shape.

Example (left): The ***circle*** circumscribes the ***triangle*** and the ***square*** circumscribes the circle.

Class boundaries

The class boundaries are upper and lower values of a ***class interval*** for grouped ***data***.

Example: For the class 5 cm $\leqslant x <$ 10 cm, the class boundaries are 5 cm and 10 cm.

Class interval

A class interval is one of the groups into which a set of numerical ***data*** has been grouped.

Class limits

These are the upper and lower values of a ***class interval***. They have the same meaning as ***class boundaries***.

Class size

The class size (also called the class width) is the difference between the upper and lower values of the class.

Example: For the class 15 cm $\leqslant x$ cm $<$ 20 cm, the class size is 20 cm $-$ 15 cm $=$ 5 cm.

Clockwise see Angles

Closed

A ***set*** is closed under an operation (for example addition) when any two members combined by that operation give another member of the set.

Examples:

The set of positive integers is closed under addition because the sum of any two positive integers is another positive integer.

The set of positive integers under division is NOT closed because, for example $3 \div 2 = 1.5$ and 1.5 is NOT a member of the set of positive integers.

$$\left\{\begin{pmatrix}1 & 0\\0 & 1\end{pmatrix}, \begin{pmatrix}-1 & 0\\0 & 1\end{pmatrix}, \begin{pmatrix}1 & 0\\0 & -1\end{pmatrix}, \begin{pmatrix}-1 & 0\\0 & -1\end{pmatrix}\right\}$$

under multiplication is closed, because the ***product*** of any two ***members*** of the set is in the set.

{1, 2, 3, 4, 5, 6, 7, 8} under addition is NOT closed because, for example, $4 + 5 = 9$ and 9 is NOT in the set.

Coefficient

In a mathematical ***expression*** the coefficients are the numbers of each ***variable***, or combination of variables.

Example: In $5x + 3y$, 5 is called the coefficient of x, and 3 is called the coefficient of y.
In $2x^2 + 4xy$, 2 is the coefficient of x^2 and 4 is the coefficient of xy.

Collinear

Collinear points lie on the same straight ***line***.

Example: These points are collinear.

Column vector

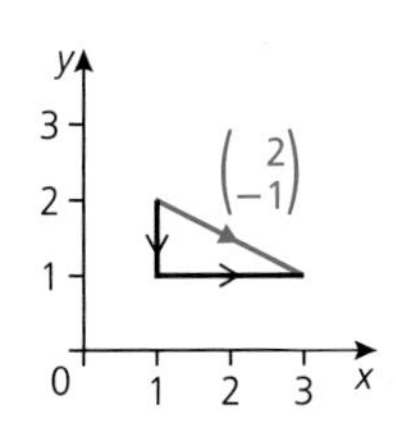

Vectors can be written as a column.
When there are two numbers, the top number gives the ***displacement parallel*** to the x-axis and the lower number gives the displacement parallel to the y-axis.

Commutative

An ***operation*** on a ***set*** is commutative when it does not matter in which order the elements are combined. The operation $*$ on a set is commutative if

$a*b = b*a$

for all ***members*** of the set.

Examples:

addition of numbers is commutative $3 + 5 = 5 + 3$

and

multiplication of numbers is commutative $7 \times 2 = 2 \times 7$

but,

division of numbers is NOT commutative $3 \div 6 \neq 6 \div 3$

and

subtraction of numbers is NOT commutative $4 - 2 \neq 2 - 4$

Addition of ***matrices*** is commutative

$$\begin{pmatrix} 0 & 1 \\ 2 & 3 \end{pmatrix} + \begin{pmatrix} 4 & 7 \\ 0 & 1 \end{pmatrix} = \begin{pmatrix} 4 & 7 \\ 0 & 1 \end{pmatrix} + \begin{pmatrix} 0 & 1 \\ 2 & 3 \end{pmatrix}$$

but multiplication of matrices is NOT commutative

$$\begin{pmatrix} 1 & -1 \\ 2 & 1 \end{pmatrix} \begin{pmatrix} 3 & 0 \\ 1 & -1 \end{pmatrix} \neq \begin{pmatrix} 3 & 0 \\ 1 & -1 \end{pmatrix} \begin{pmatrix} 1 & -1 \\ 2 & 1 \end{pmatrix}$$

Compass direction

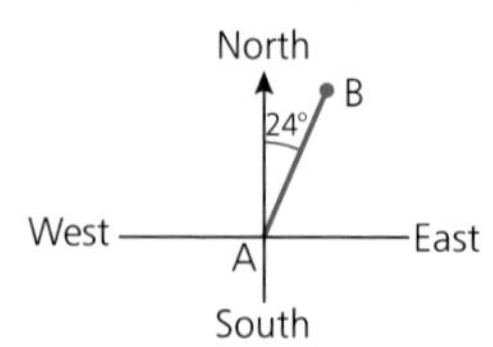

Compass directions are given as ***angles*** north or south of the main compass points of North, East, South or West.
See **Bearing**.

Example: Point B is in the direction N 24° E from point A.

Complement of a set

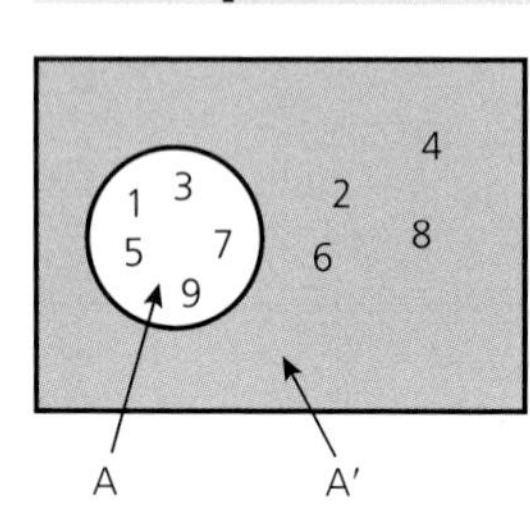

The complement of a ***set*** A in U is the set of all ***elements*** of U NOT in A, and is shown by the ***symbol*** A′.

Example (left):
If $U = \{1, 2, 3, 4, 5, 6, 7, 8, 9\}$
and A is the {***odd numbers***} = {1, 3, 5, 7, 9}
then A′ is the {***even numbers***} = {2, 4, 6, 8}
The complement of A is the region shaded green.
NOTE: A ***union*** A′ is the ***universal set*** $A \cup A' = U$

Complementary angles

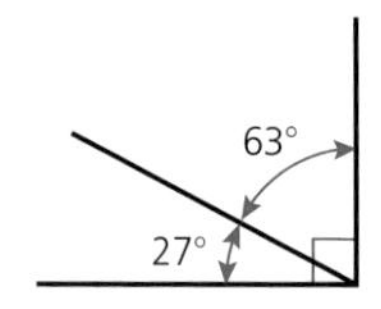

If two ***angles*** such as 63° and 27° add up to 90° they are called complementary angles and one is said to be the complement of the other.

$27° + 63° = 90°$

Completing the square

Completing the square means adding a number to a ***quadratic expression*** to make a ***perfect square***.

Example: To complete the square for $x^2 + 6x$ we need to add 9.
Then $x^2 + 6x + 9 = (x + 3)^2$

Composite number

A composite number is a ***whole number*** that has ***factors*** other than one and itself.
Example: 12 is a composite number because $12 = 3 \times 4$.

Compound interest

When interest is paid on a sum of money that includes interest from previous interest payments, the compound interest is the difference between the ***principal*** and the ***amount***.

Compound proportion

Compound proportion involves more than two quantities that are in proportion.

Example: Water, sugar and lemon juice are mixed in the ratio 8:1:2 to make a drink.

Cone

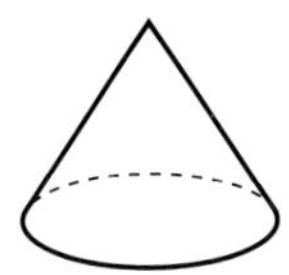

A cone is a solid which usually has a circular base and tapers to a point at the top, called its ***vertex***.

Examples:

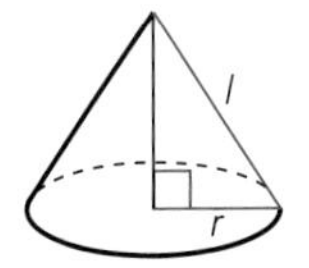

The ***area*** of the curved part of a cone is given by
curved area = $\pi \times$ ***radius*** $\times$ ***slant height***
$A = \pi r l$

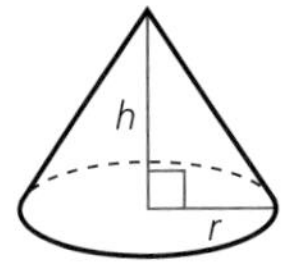

The ***volume*** of a cone is given by
volume = $\frac{1}{3} \times$ base area $\times$ height
$V = \frac{1}{3}\pi r^2 h$

Congruent

If two objects have the same shape and size they are congruent.

Examples: Each cup and saucer (left) is congruent to every other cup and saucer.

 and are congruent shapes.

 and  are congruent shapes.

 and 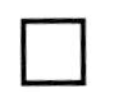 are not the same size and so they are NOT congruent.

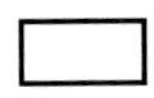 and  are not the same shape and so they are NOT congruent.

Constant

A constant is a fixed number. Its value does not change.

Example: In the ***formula*** $s = 6t + 4$ the numbers 6 and 4 are constant, in contrast to s and t which can take different number values.

Construction

A construction is an accurate drawing of a ***plane*** shape.

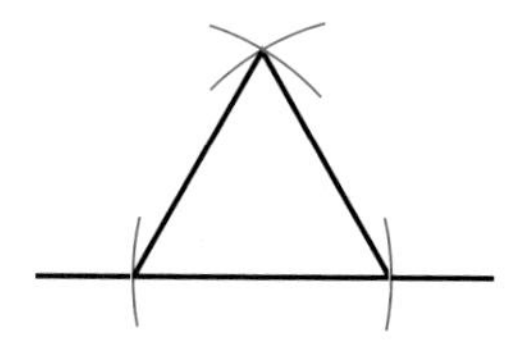

Continuous

How much do the apples weigh?

Weight is a continuous quantity, because the answer can be ANY value: 2.1841 … kg
or 2.1842 … kg

The spring balance measures a change in weight continuously but a scale pan using individual weights can only measure changes equal to the smallest weight.

Other continuous quantities are length, ***area***, height, ***volume***, temperature, and ***speed***. As a car ***accelerates*** from start to 50 miles per hour, the speed must pass through ALL values between 0 and 50.

Quantities which are NOT continuous are called ***discrete***.

NOTE: When x is continuous we use ALL the points on the ***number line***.

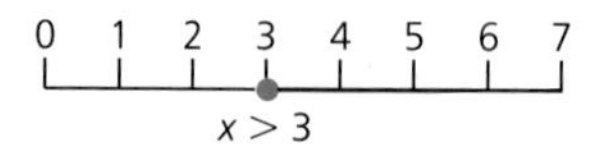

Convex

These *polygons* are called convex because all their *vertices* point outwards.

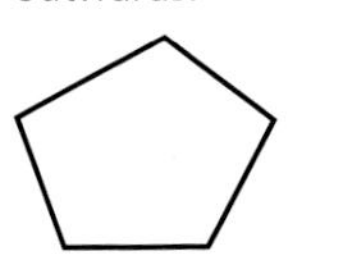

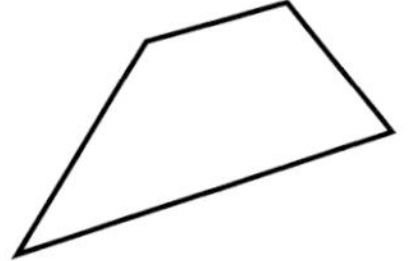

A shape is convex if the *line* joining ANY two points on the shape stays inside the shape.

Shapes which are NOT convex are called *re-entrant*.

Coordinates

A point on a *graph* can be fixed by a pair of numbers which describe its position with reference to the x and y *axes*.

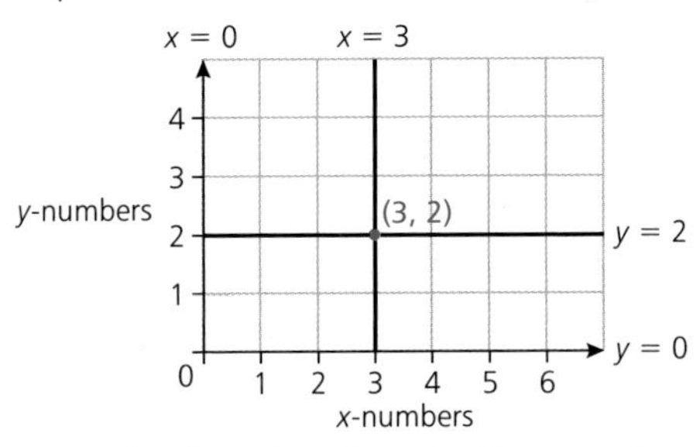

Example: The point where the *lines* $x = 3$ and $y = 2$ cross is labelled (3, 2).

The *coordinates* of the point are 3 and 2.

The first number, 3, is called the x-coordinate.

The second number, 2, is called the y-coordinate.

NOTE: (2, 3) is NOT the same point as (3, 2). See **Ordered pairs, Cartesian plane**

Correspondence

Arrow diagrams are classified into four types (called correspondences).

Examples:

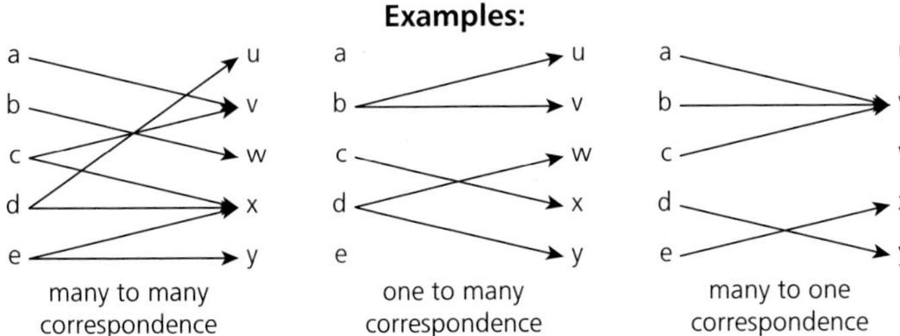

many to many correspondence

one to many correspondence

many to one correspondence

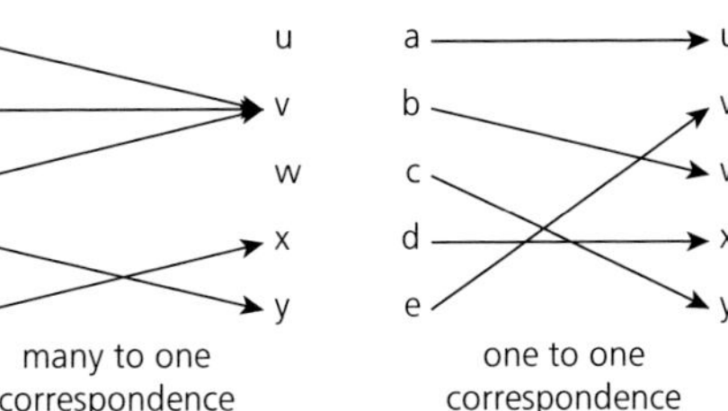

one to one correspondence

These *relations* are NOT *mappings*.

These relations ARE mappings.

Corresponding points

Points are said to be corresponding when, under ***transformations*** such as ***enlargement***, ***rotation***, ***reflection***, ***translation*** etc., one point is the ***object*** and the other is its ***image***.

Examples:

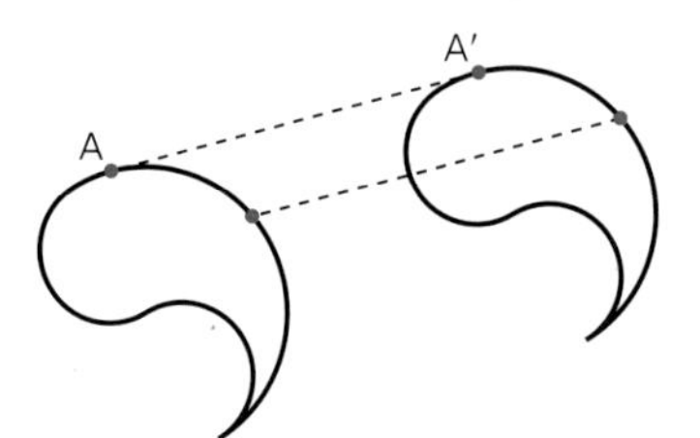

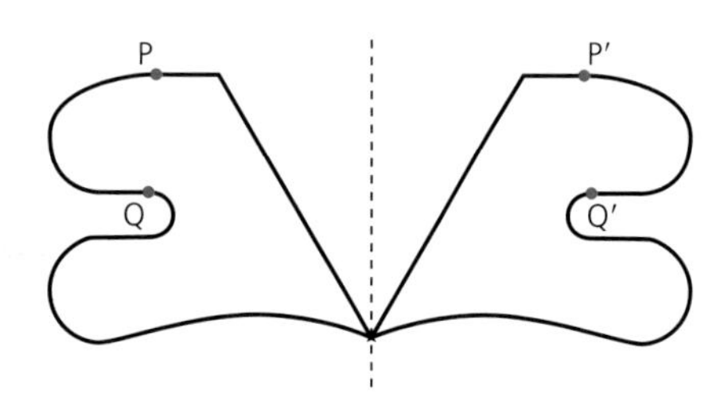

A and A′, P and P′, Q and Q′ are corresponding points.

Cosine

The cosine of ***angle*** θ is written as $\cos \theta$

Cosine of angles less than 90°

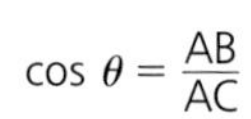

$$\cos \theta = \frac{AB}{AC} \qquad \cos \theta = \frac{\text{adjacent side}}{\text{hypotenuse}}$$

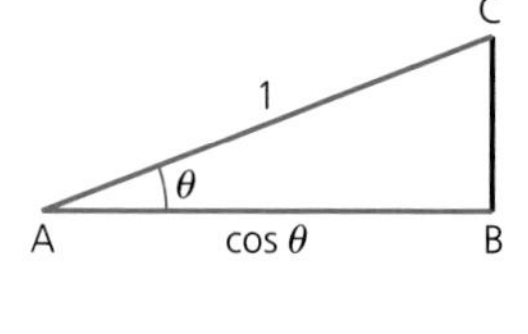

When the length of AC is 1 then

$\cos \theta = AB$

When this triangle is ***enlarged***, with ***scale factor*** *r*.

$PQ = r \cos \theta$

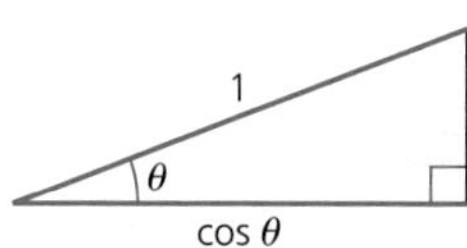

S.F. × *r*

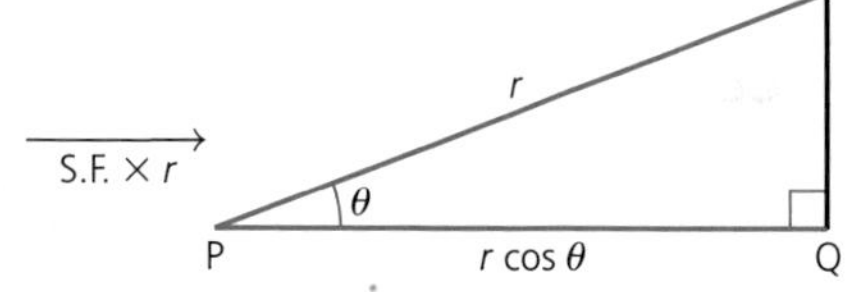

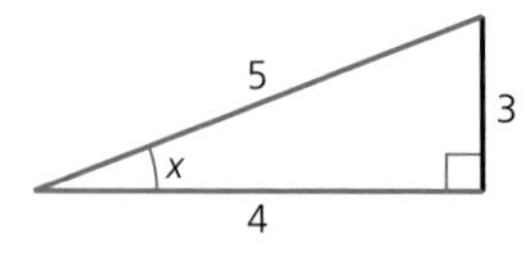

Examples:

$\cos x = \frac{4}{5}$

$\cos x = 0.8$

$\cos 36.9° = 0.800$

$x = 36.9°$

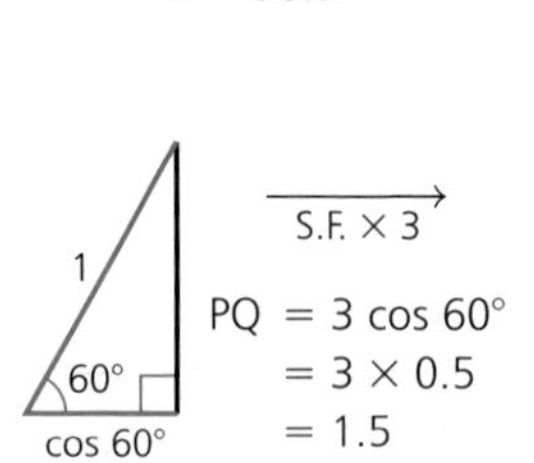

S.F. × 3

$PQ = 3 \cos 60°$

$= 3 \times 0.5$

$= 1.5$

3, 60°, P, 3 cos 60°, Q

$\cos 60° = 0.5$

Cosine of any angle

If a ***line*** OP of unit length turns through an angle θ from the ***x-axis***; then the cosine of θ is the *x*-***coordinate*** of P shown in green.

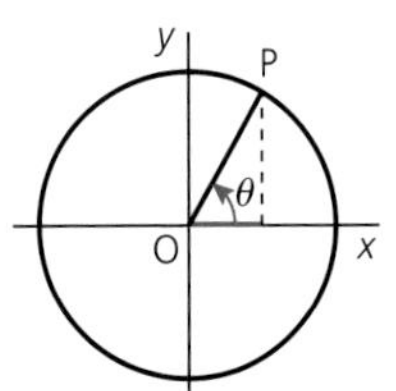

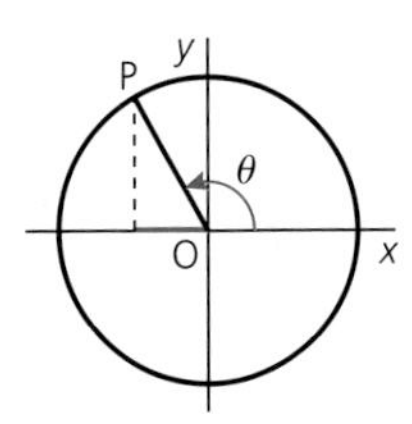

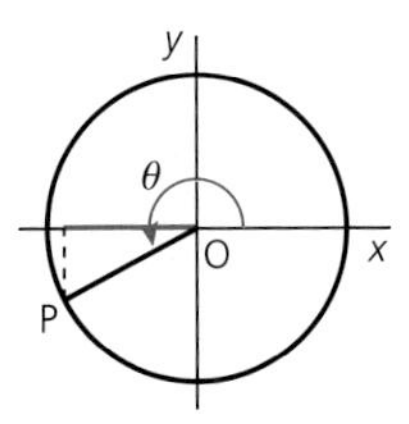

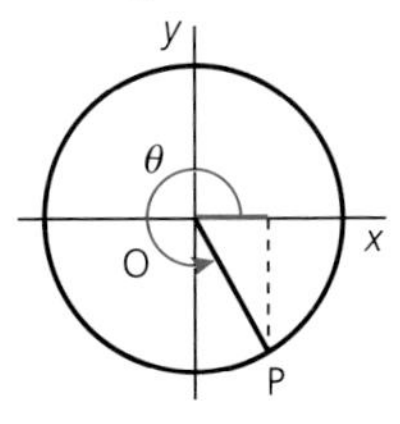

The graph of $f(x) = \cos x$ is

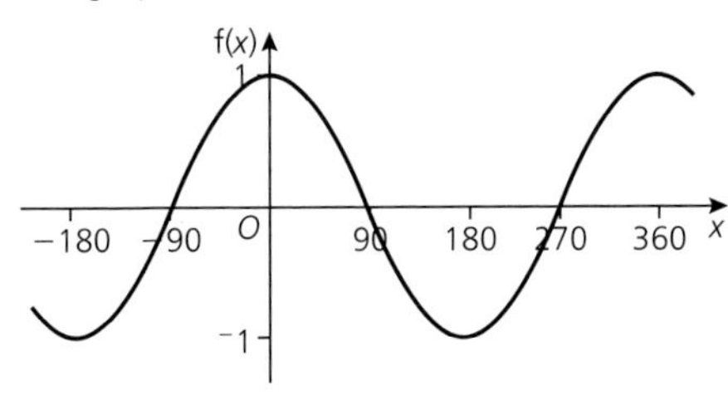

Cosine formula

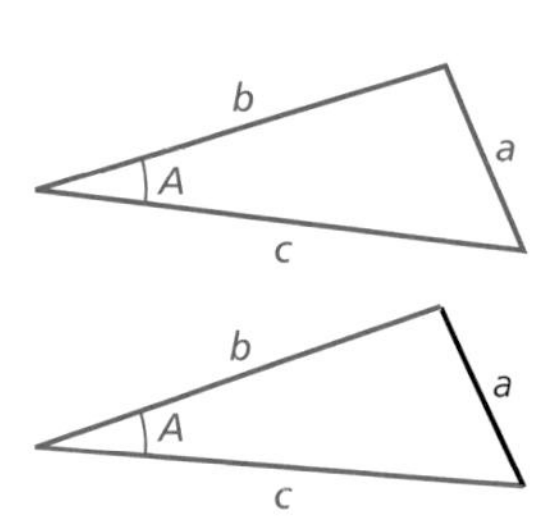

When the lengths of the sides of a ***triangle*** are known, an angle, A, can be found by the formula

$$\cos A = \frac{b^2 + c^2 - a^2}{2bc}$$

When an angle and the two sides which form the angle are known, the third side can be found using the formula in this form

$$a^2 = b^2 + c^2 - 2bc \cos A$$

Counting numbers

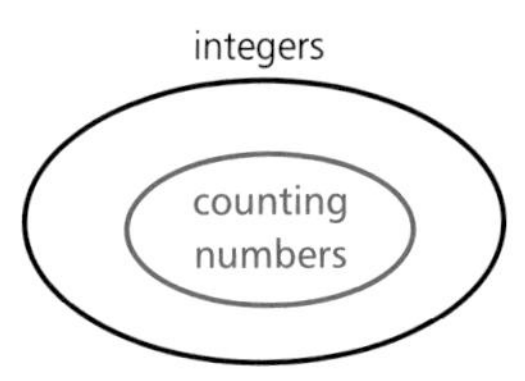

The ***set*** of counting numbers is the set of numbers people first used to count things such as sheep.

The set of counting numbers is $\{1, 2, 3, 4, 5, 6, 7, 8, \ldots\}$

The set of counting numbers is the same as the set of ***natural numbers*** and is a ***subset*** of the ***integers***.

Cube

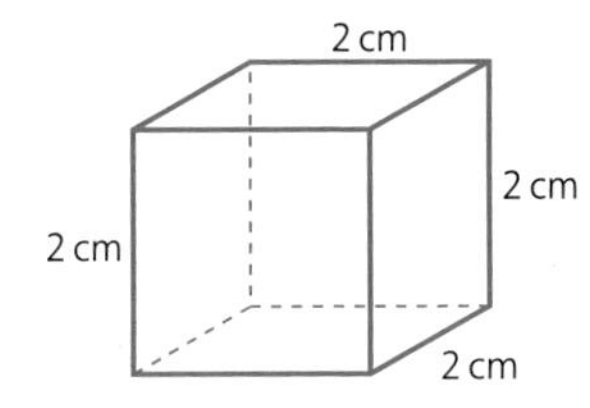

A cube is a ***regular*** solid with all its ***faces square*** and all its edges equal in length. The shape of a die is a cube.

This is a diagram of a 2 cm cube.

Examples:

Cube root

When a number is cubed the starting number is the cube root.

Example: $2 \times 2 \times 2 = 8$ so 2 is the cube root of 8.

Cubed

A number 'cubed' is the number to the ***power*** 3.

Example: 2 cubed is $2^3 = 8$

Cubic centimetre

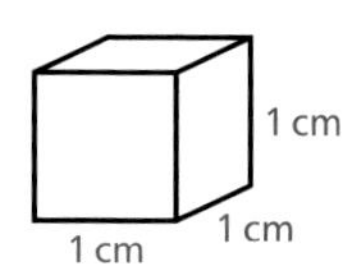

A cubic centimetre is a ***unit*** for measuring ***volume***. Cubic centimetre is shortened to cm^3.

One cubic centimetre is the same volume as a ***cube*** whose edges are one ***centimetre*** in length.

Example: The volume of this cuboid is $2 \times 3 \times 7 = 42$ cm^3

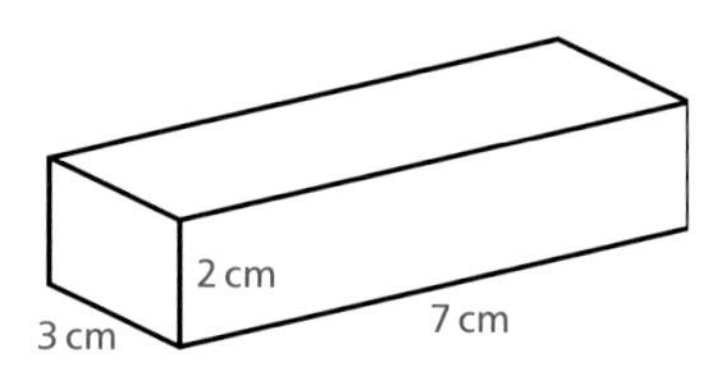

Cubic metre

A cubic metre is the standard unit for measuring ***volume***. Cubic metre is shortened to m^3.

One cubic metre is the same volume as a ***cube*** whose edges are one ***metre*** in length.

Example: The volume of this *cuboid* is $\frac{1}{2} \times \frac{1}{2} \times 20 = 5$ m^3

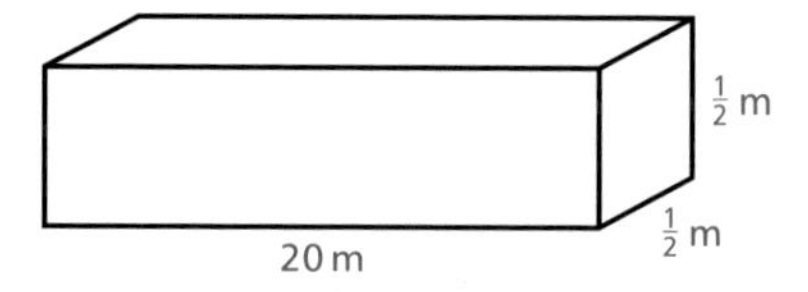

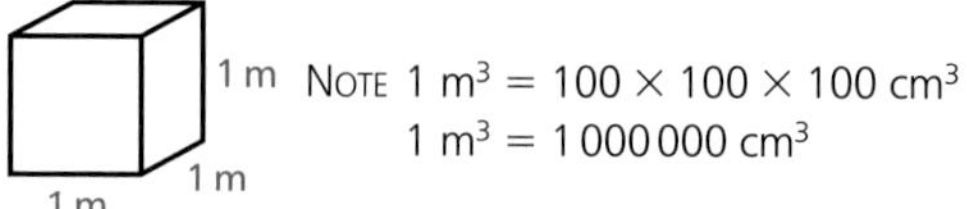

NOTE $1\ m^3 = 100 \times 100 \times 100\ cm^3$

$1\ m^3 = 1\,000\,000\ cm^3$

Cuboid

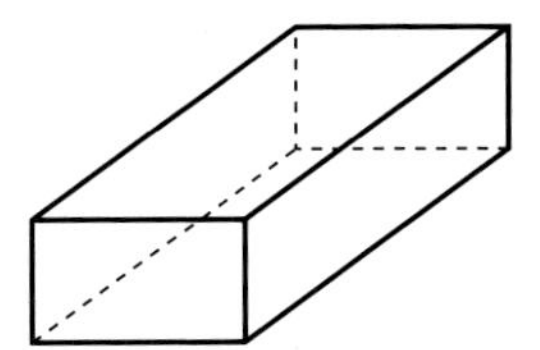

A cuboid is a solid which has ***rectangles*** for all of its ***faces***.

Examples:

Cumulative frequency

The cumulative frequency is the ***sum*** of the ***frequencies*** at or below a given value.

Example: In a survey 100 children were asked what size of shoe they wore.

This is a bar chart.

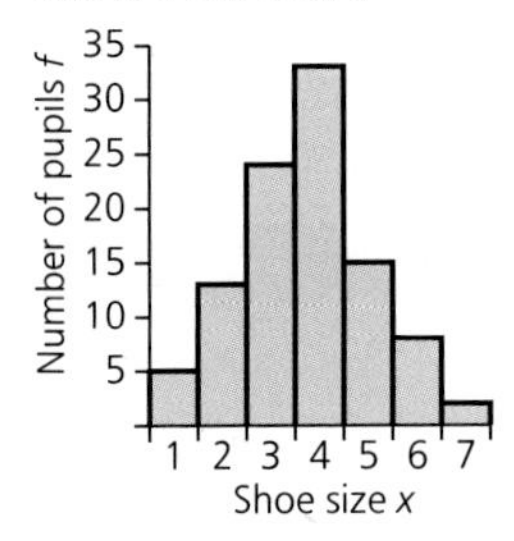

This is a cumulative frequency table.

Shoe size x	f frequency	cumulative frequency
0	0	0
1	5	5
2	13	18
3	24	42
4	33	75
5	15	90
6	8	98
7	2	100

This is a cumulative frequency curve, or ***ogive***.

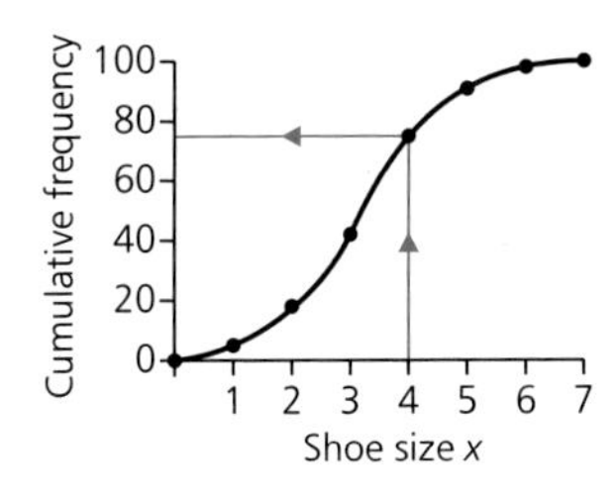

The frequency of shoe size 4 is 33 but there are 75 children with a shoe size of 4 or less. 75 is the cumulative frequency of size 4.

Cyclic quadrilateral

A cyclic quadrilateral has all four ***vertices*** on a ***circle***.

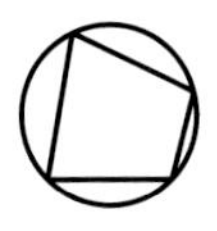

Cylinder

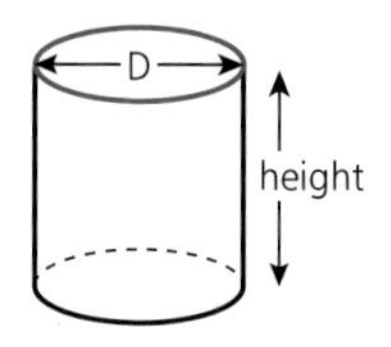

A cylinder is a ***prism*** with the shape of a ***circle*** along its length.

Examples:

The ***area*** of the curved part of the cylinder is given by
curved area = ***circumference*** × height

This can be opened into this

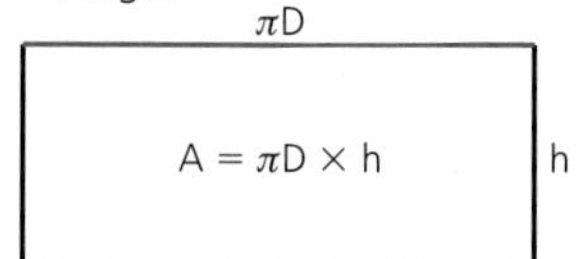

The ***volume*** of the cylinder is given by
volume = area of end × height

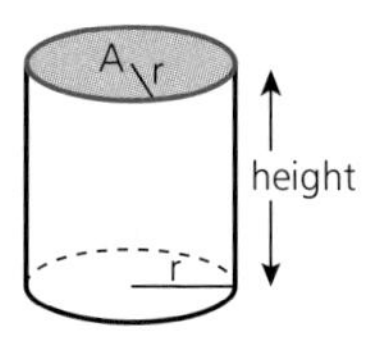

$V = A \times h$
$V = \pi r^2 h$

D

Data

Data are numbers which have been collected for study. Data can be displayed as ***bar charts***, ***pie charts***, ***pictograms***, etc.

Example: A class of 23 pupils are asked how many children there are in each of their families. The following data are collected 2, 1, 1, 2, 3, 3, 1, 2, 1, 4, 1, 2, 2, 3, 3, 3, 6, 4, 2, 3, 2, 2, 3.

There are five families with 1 child, eight families with 2 children, seven families with 3 children, two families with 4 children and just one family with 6 children.

Decagon

A decagon is a ***polygon*** bounded by ten straight lines and with ten ***angles***.

Examples:

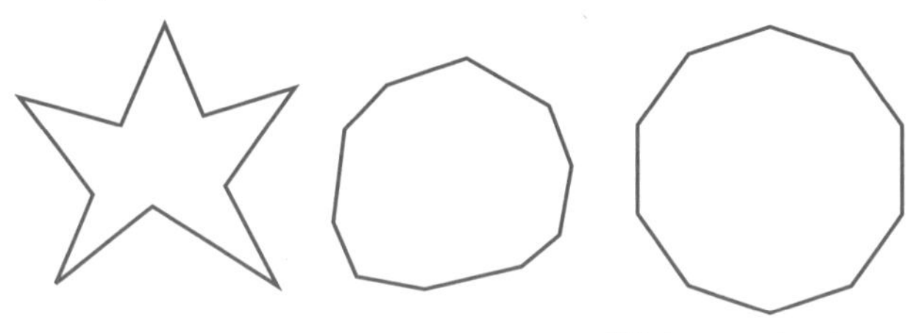

This is a ***regular*** decagon

Decimal

A decimal is a number written in ***base*** 10. A point separates the whole number from the fractional part. The position of a ***digit*** gives the ***place value*** as a ***power*** of ten.

Example: In 36.27, the digit 3 represents 3×10^1 (3 tens),
the digit 6 represents 6×10^0 (6 units),
the digit 2 represents 2×10^{-1} (2 tenths),
the digit 7 represents 7×10^{-2} (7 hundredths).

Decimal place

The decimal place describes the position of a digit to the right of the ***decimal*** point.

Example: In 36.27, 2 is in the first decimal place and 7 is in the second decimal place.

Definite integral

A definite integral is a number equal to the value of the integral at an upper value of the variable minus the value of the integral at a lower value of the variable.

Example: $\int_2^4 2x\mathrm{d}x = \left[x^2\right]_2^4 = 4^2 - 2^2 = 16 - 4 = 12$

See **Integration**, **Indefinite integral**

Denominator see Fraction

Density

Density is the ***mass*** of one ***unit*** of ***volume*** of a material. It is usually measured in ***grams*** per ***cubic centimetre***, g/cm^3.

Depression

The ***angle*** of depression is the angle measured DOWNWARDS from the ***horizontal*** to an object.

Example:

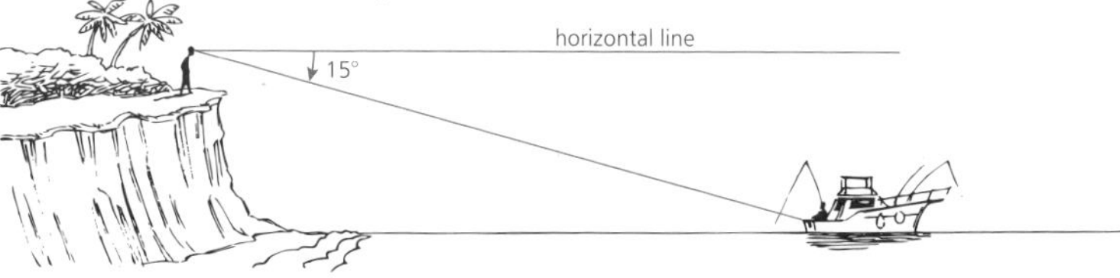

The angle of depression of the boat from the man is 15°.

Derivative

The derivative of a ***function***, f(x), is the ***rate of change*** of f(x) with respect to x and is equal to limit as $h \to 0$ of $\frac{f(x+h) - f(x)}{h}$ where h is a small increase in x.

The result is the derived function, $f'(x)$.

Determinant

The determinant of a ***matrix***, **M**, is a special number that goes with matrix, **M**. It can be found like this:

when $\mathbf{M} = \begin{pmatrix} a & b \\ c & d \end{pmatrix}$

then the determinant of **M** is ad − bc.

'The determinant of matrix **M**' is shortened to $|\mathbf{M}|$ or det **M**.

Example:

$$\det \begin{pmatrix} 3 & 5 \\ 2 & 4 \end{pmatrix} = \begin{vmatrix} 3 & 5 \\ 2 & 4 \end{vmatrix}$$

$$= 3 \times 4 - 2 \times 5$$

$$= 2$$

When the matrix is used to represent a ***transformation*** its determinant is the area ***scale factor*** of the transformation.

Diagonal

A diagonal is a straight ***line*** drawn from one ***vertex*** of a ***polygon***, or a ***polyhedron***, to another vertex. The side of a polygon is not a diagonal; and an ***edge*** of a polyhedron is not a diagonal.

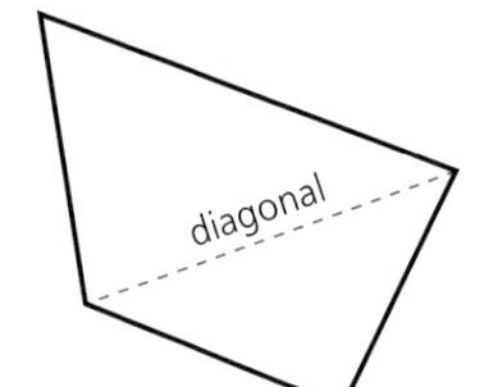

Examples:

Diagonal of a polygon

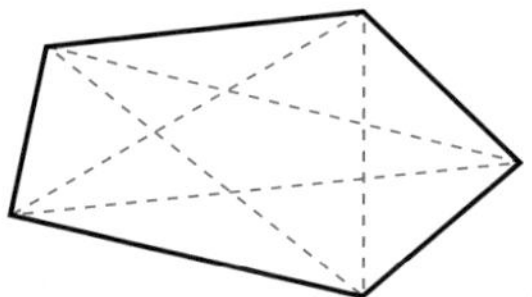

All the diagonals are shown on this figure.

Diagonal of a polyhedron

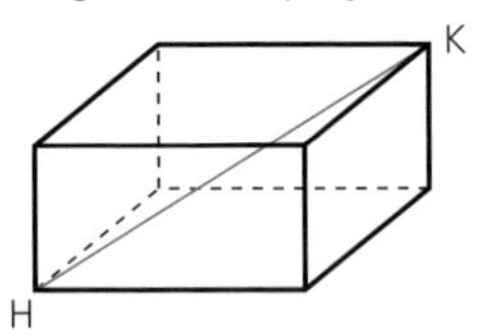

The one and only diagonal from the vertex H is shown by the green line HK.

Diameter

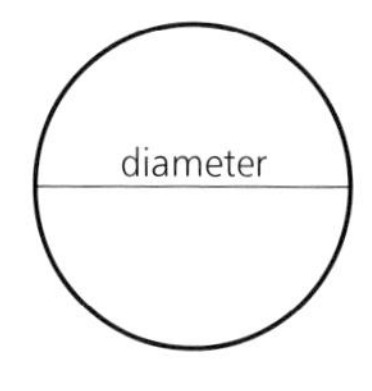

A diameter of a ***circle*** is any ***line*** that joins two points of the circle and passes THROUGH THE CENTRE.

Examples:

All the diameters of a circle are the same length, twice the ***radius***.

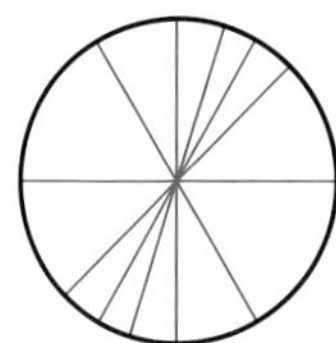

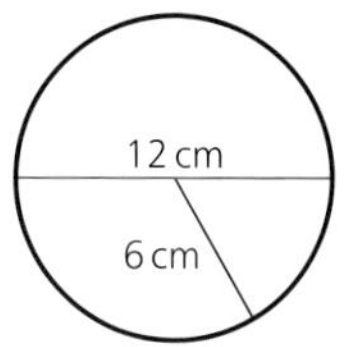

Difference

The difference of two numbers is the answer when the smaller number is ***subtracted*** from the bigger number.

Examples:

The difference between 8 and 3 is 5: $8 - 3 = 5$
and
the difference between 3 and 8 is 5: $8 - 3 = 5$.

Differentiation

Differentiation is the operation of finding the ***derivative*** of a ***function***.

Digit

A digit is any single figure used when representing a number.
In our ***base*** ten number system that we use daily the digits are 0, 1, 2, 3, 4, 5, 6, 7, 8, 9.

Example: The digits used in 107585 are 0, 1, 5, 7, 8.

Direct proportion

Two quantities are in direct proportion when one is always a ***multiple*** of the other.

Example: When a bus travels at a constant speed of 50 km/h, the distance covered by the bus is 50 times the number of hours it takes. So the distance is directly proportional to the time taken.

See **Direct variation**

Direct variation

Direct variation is the same as ***direct proportion*** but is usually used for two ***variables***.

If x and y are two variables, y varies directly as x when $y = kx$ where k is a ***constant***.

Dividend

In a division the dividend is the number being divided.

Example:
In the division $4\overline{)20812}$ with quotient 5203, 20 812 is the dividend.

See **Quotient**

Divisor

In a division, the divisor is the number you are dividing by.

Example:
In the division $3\overline{)6042}$ with quotient 2014, the divisor is 3

Dodecagon

A ***polygon*** bounded by twelve straight ***lines*** and containing twelve ***angles*** is a dodecagon.

Examples:

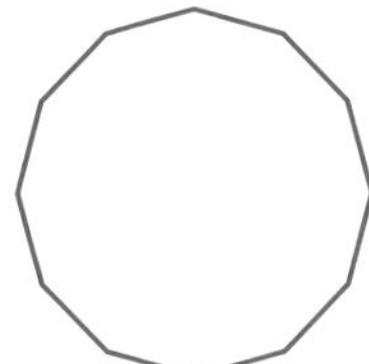

This is a ***regular*** dodecagon.

Dodecahedron

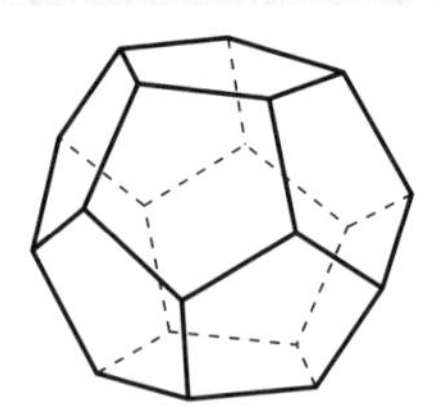

A dodecahedron is a solid shape with twelve ***faces***. All the faces of a ***regular*** dodecahedron are regular ***pentagons***.

Example (left): This is a regular dodecahedron.

Domain

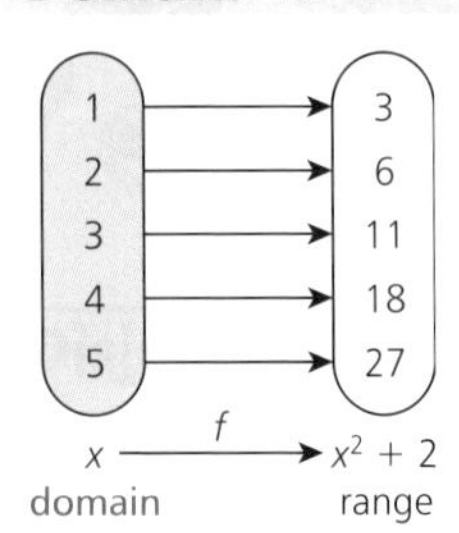

The domain of a ***function*** is the ***set*** of numbers ***mapped*** by the function.

NOTE: The domain is sometimes called the ***object*** set.

Example (left): {1, 2, 3, 4, 5} is the domain.

See **Range**

Duodecimal

$t_{twelve} = 10_{ten}$
$14_{twelve} = 16_{ten}$
$18_{twelve} = 20_{ten}$
$21_{twelve} = 25_{ten}$

A duodecimal number is a number written in ***base*** twelve.

If we use the symbols t for ten and e for eleven then the ***digits*** in base twelve are 0, 1, 2, 3, 4, 5, 6, 7, 8, 9, t, e

E

Edge

The edge of a ***polyhedron*** is where two ***faces*** of the polyhedron meet.

Element

The elements of a ***set*** are the things in the set.

Examples:
If A = {first 5 letters of the alphabet}, then the elements of A are a, b, c, d, and e.

If P = {prime numbers between 4 and 20}, then the elements of P are 5, 7, 11, 13, 17, and 19.

The number of elements in set A is five, this is written as

$$n(\mathrm{A}) = 5$$

Similarly, $n(\mathrm{P}) = 6$

Each letter or number in a ***matrix*** is also called an element.

Elevation

The ***angle*** of elevation is the angle measured UPWARDS from the ***horizontal*** to an object.

Example:

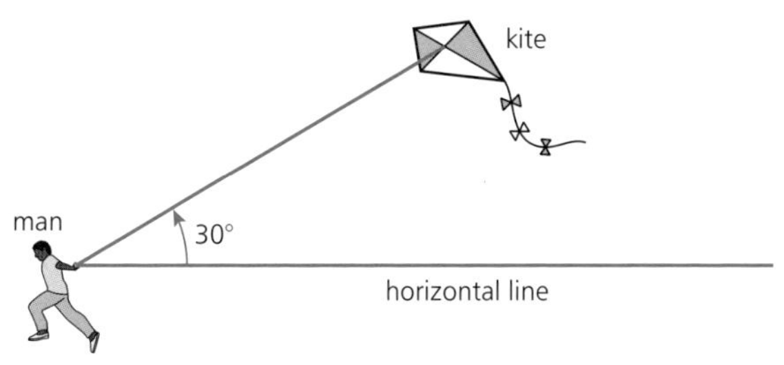

The angle of elevation of the kite from the man is 30°.

Empty set

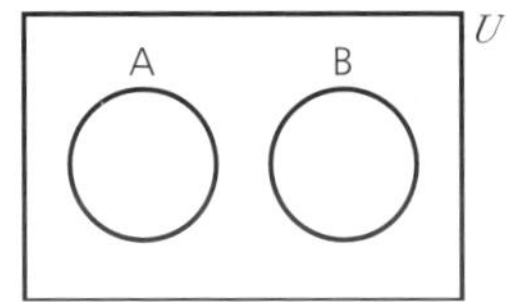

When a ***set*** has no ***elements*** it is called the empty set and is shown as { }, or by the symbol ∅

Example (left): If A = {6, 8, 10, 12} and B = {1, 3, 5, 7) then A ∩ B = { } or A ∩ B = ∅

The empty set is called the null set.

Enlargement

Enlargement is a ***transformation*** which ***maps*** a shape onto a ***similar*** shape, from a centre O using a ***scale factor*** *k*.

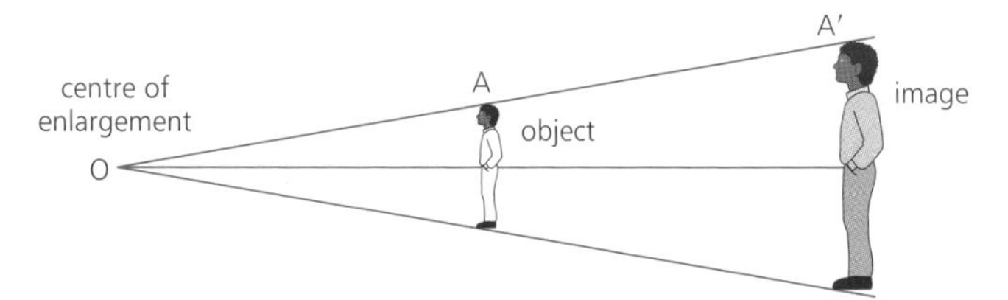

To find the ***image***, A′, of a point A of the ***object***, a line is drawn from O through A. A′ is then on OA and OA′ = $k \times$ OA
In the example above the scale factor is 2.

The lines through O used to draw the image are called ***guide lines***. When the scale factor *k* is a ***fraction*** then the image is smaller than the object.

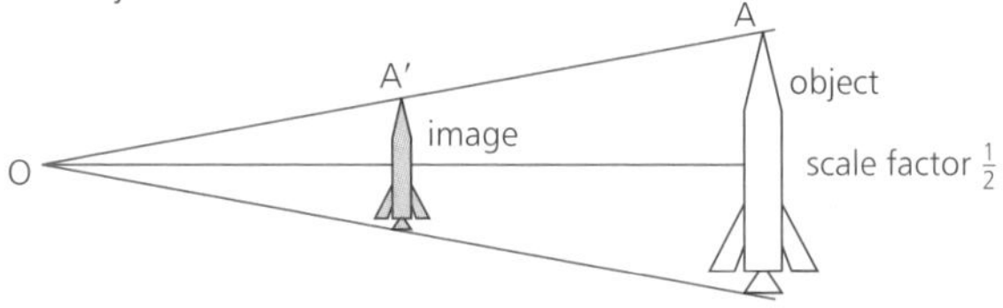

When the scale factor *k* is ***negative*** then the enlarged image is also ***rotated*** 180° about O.

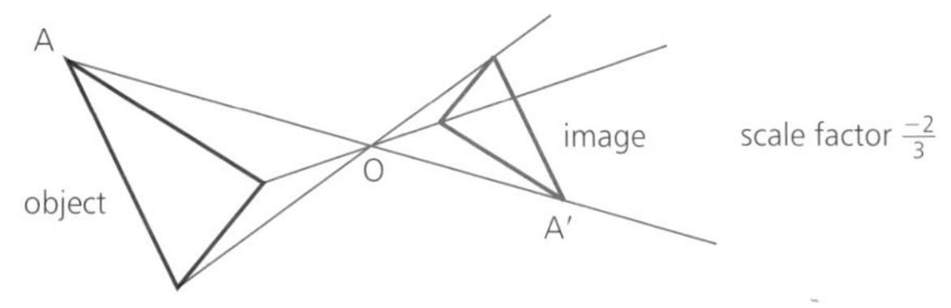

Enlargement matrix

The ***transformation*** $\begin{pmatrix} x \\ y \end{pmatrix} \rightarrow \begin{pmatrix} k & 0 \\ 0 & k \end{pmatrix} \begin{pmatrix} x \\ y \end{pmatrix}$ represents an enlargement with centre at the ***origin***, and scale factor *k*.

NOTE: This scale factor, *k*, is the linear scale factor. The area scale factor is k^2, the ***determinant*** of the ***matrix***.

Equal set

Two ***sets*** are equal when they have the same ***elements***.

Example: The sets {A, B, C, D} and {D, B, A, C} are equal sets.
See and compare with **Equivalent set**

Equally likely

Events are equally likely if the chance of them happening is the same.

Examples: If I roll a die, each of the six numbers is equally likely to show on top.

If I roll two dice and add the top numbers there are eleven possible totals, 2, 3, 4, 5, 6, 7, 8, 9, 10, 11, 12, and they are NOT equally likely. A 2 will only happen if I roll 1 and 1, but a total of 7 will come from any of the pairs 1 and 6, 2 and 5, 3 and 4, 4 and 3, 5 and 2, 6 and 1.
See **Probability**

Equation

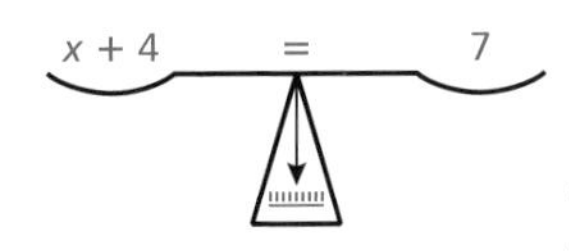

An equation is a statement using an equals sign.

Examples:

$y = 3x \qquad x + 4 = 7$

$y = x^2 - 5 \qquad 2x + 5 = 7x$

See **Expression**

Relations as an equation

An equation is one way of showing a ***relation***.

Examples:

The ***linear function*** $x \rightarrow x + 3$

can be written as the ***linear equation*** $y = x + 3$

The ***quadratic function*** $x \rightarrow x^2 + 2$

can be written as the equation $y = x^2 + 2$

See **Equation of a line, Quadratic equation, Simultaneous equations, Solution of an equation**

Equation of a circle

The equation of a circle is the relationship between the ***coordinates*** that is satisfied by all points on the ***circle***.

The general ***equation*** of a ***circle*** is $(x - a)^2 + (y - b)^2 = r^2$ where (a, b) is the centre of the circle and r is the ***radius***.

Equation of a line

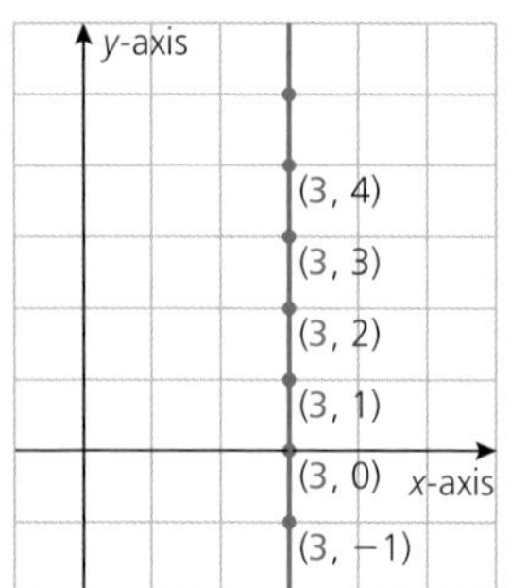

The equation of a line is the ***relation*** satisfied by the ***coordinates*** of the points of the ***line***.

As the first coordinate, x, is 3 for every point on the line, the ***equation*** of this line is $x = 3$.

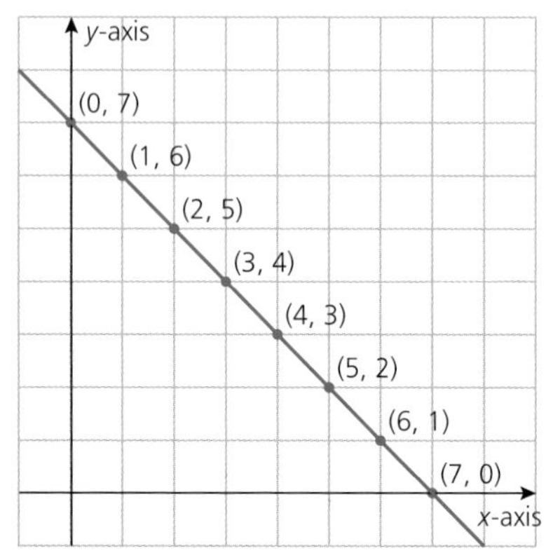

As the two coordinates always add up to 7 for the points on this line, its equation is $x + y = 7$

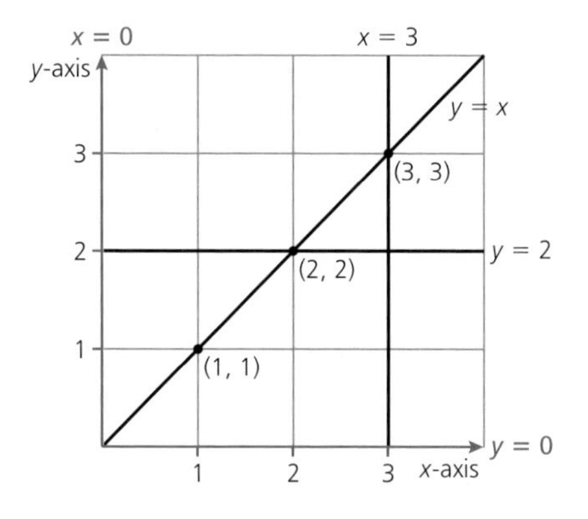

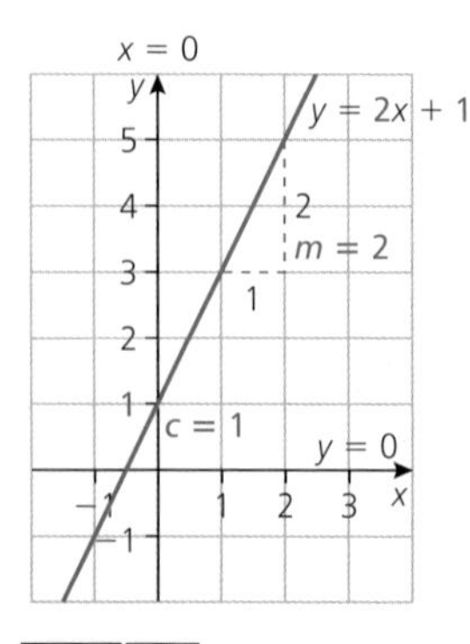

NOTE: The equation of the *x-**axis*** is $y = 0$ and the equation of the y-axis is $x = 0$.

Any straight line on a ***graph*** represents a ***linear equation*** which can be written in the form $y = mx + c$

Examples:

$y = 3x - 4$ $\quad y = x + 1$

$y = 8 - x$ $\quad y = 2 - \frac{1}{2}x$

When a linear equation is written in the form $y = mx + c$ the number m represents the ***gradient*** of the line and the number c represents the intercept on the y-axis.

Equilateral triangle

An equilateral triangle is a ***triangle*** with all three sides the same length.

Examples:

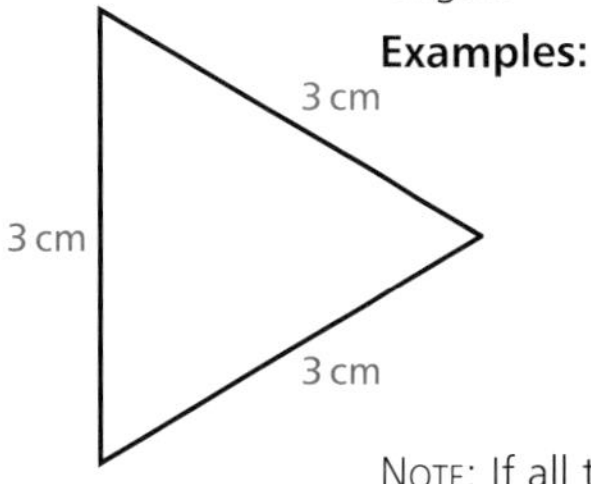

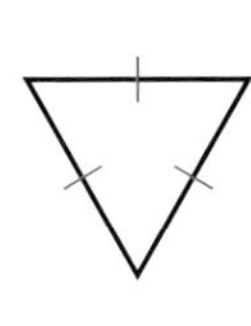

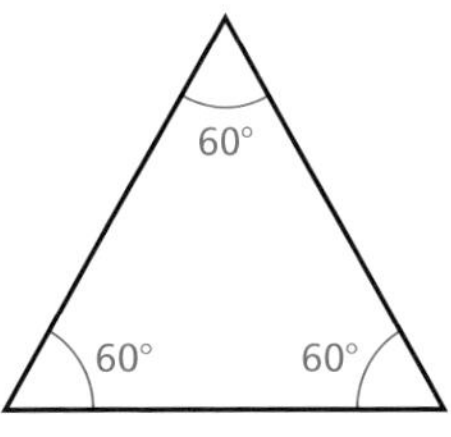

NOTE: If all three sides of a triangle are equal, the ***angles*** will be equal and will be 60°.

Equivalent fraction

Fractions are equivalent if they can be ***cancelled*** to the same fraction.

Example: $\frac{8}{12}$ and $\frac{6}{9}$ are equivalent because they can both be cancelled to $\frac{2}{3}$.
Here are two ***sets*** of equivalent fractions:

$\left\{\frac{1}{2}, \frac{2}{4}, \frac{4}{8}, \frac{5}{10}, \frac{8}{16}, \frac{10}{20}\right\}$

$\left\{\frac{5}{6}, \frac{50}{60}, \frac{10}{12}, \frac{25}{30}, \frac{500}{600}\right\}$

If $\frac{a}{b}$ is any fraction then $\frac{ka}{kb}$ is an equivalent fraction where k is a number not equal to ***zero***.

Equivalent set

Two ***sets*** are equivalent when they have the same number of ***elements***.

Example: The sets {A, B, C, D} and {1, 2, 3, 4} are equivalent because they both have four elements.
See and compare with **Equal set**

Error

An error is the difference between an ***estimated*** value and the actual value of a quantity.
See **Absolute error**, **Percentage error**, **Relative error**

Escribed circle

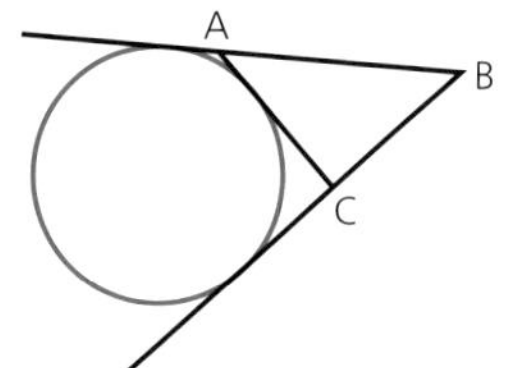

An escribed circle is a ***circle*** that touches one side of a ***triangle*** and also touches the extensions of the other two sides.
Example (left): The diagram shows one escribed circle of the triangle ABC.

Formula

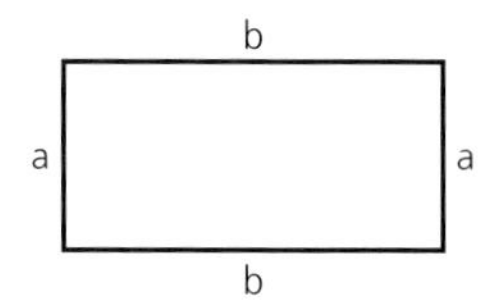

A formula is a general ***equation***. It shows the connection between related quantities.

Example: The ***perimeter*** P of a ***rectangle*** with sides a and b is given by the formula

$P = 2a + 2b$

Some examples of formulas you may meet are:

$v = u + at, \quad C = \pi d, \quad T = 2\pi\sqrt{\frac{l}{g}}, \quad \frac{1}{f} = \frac{1}{v} + \frac{1}{u},$

$A = \pi r^2$, and the formula for solving ***quadratic equations*** is

$$x = \frac{-b \pm \sqrt{b^2 - 4ac}}{2a}$$

Fraction

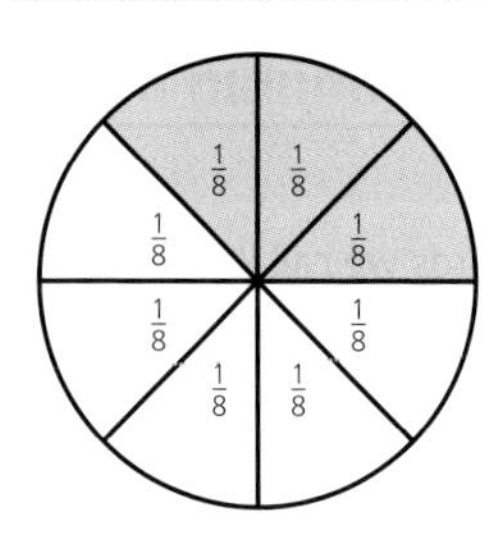

A fraction is one number divided by another number written as $\frac{a}{b}$, such as

$\frac{1}{2} \quad \frac{1}{4} \quad \frac{3}{5} \quad \frac{11}{13} \quad \frac{21}{40}$

Example (left): $\frac{\text{number of shaded parts}}{\text{number of parts of circle}} = \frac{3}{8}$

The shaded portion of the ***circle*** is $\frac{3}{8}$ of the whole circle.
NOTE: The bottom number is called the ***denominator*** and the top number is called the ***numerator***.

See **Equivalent fractions, Improper fraction**

Frequency

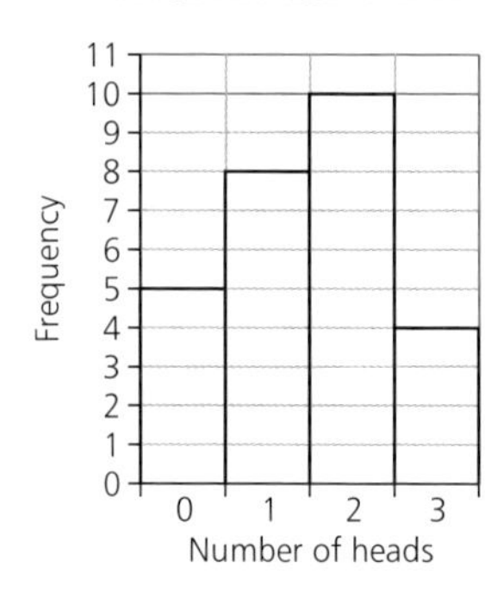

The frequency of an event is how many times it has happened.

Example: A die is rolled 100 times and a six comes up 13 times. The frequency of a score six is 13.

Frequency distribution

A frequency distribution gives the number of times that each different value or group of values occurs.

Frequency table

A frequency table shows each value or group of values and its frequency.

Example: This is a frequency table showing the distribution of the number of heads that occurred when three coins were flipped.

Number of heads	0	1	2	3
Frequency	5	8	10	4

See **Mode**

Frequency polygon

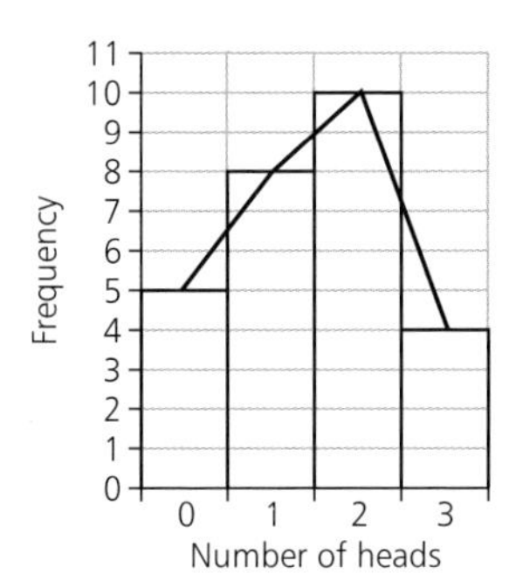

A frequency polygon illustrates a ***frequency*** distribution with a ***polygon*** drawn by joining the middle of the tops of the bars in a ***bar chart*** or a ***histogram***.

Example (left): This polygon represents the distribution frequency of heads when three coins are flipped.

Frustum

A frustum is part of a solid bounded by two ***parallel planes***. Usually one of the planes is parallel to the base.

Example: This is a frustum of a ***cone***.

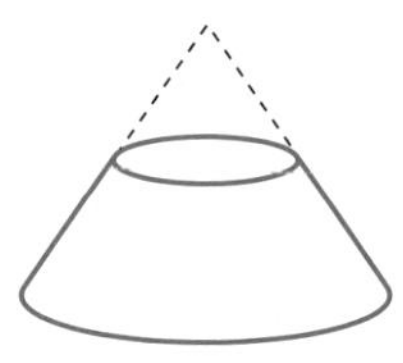

Function

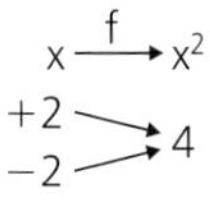

Functions are special kinds of ***relations*** in which each ***object*** is ***mapped onto*** only ONE ***image***.

They are also known as ***mappings***.

Examples:

f (*left*) is a function as every object has only ONE image.

$x \xrightarrow{g}$ square root of x

$1 \rightarrow +1$ or -1

$4 \rightarrow +2$ or -1

$9 \rightarrow +3$ or -3

However, g (*above*) is NOT a function as there is more than one image for each object.

See **Correspondence**, **Mapping**

Function notation

A function of x is written as $f : x \rightarrow 2x, x \in \mathbb{R}$ or as $f(x) = 2x, x \in \mathbb{R}$ which means x maps to $2x$ for values of x in the ***set*** of ***real numbers***.

Geometric progression

A geometric progression is a ***sequence*** of numbers where each ***term*** is a constant ***multiple*** of the term before it. The constant multiple is called the common ratio.

Example: The sequence 3, 6, 12, 24, 48, 96, … is a geometric progression. The first term is 3 and the common ratio is 2. It can be shown as $3, 3 \times 2, 3 \times 2^2, 3 \times 2^3, 3 \times 2^4, 3 \times 2^5, \ldots$

When the first term is a and the common ratio is r, the sequence is $a, ar, ar^2, ar^3, ar^4, \ldots$

The sum of the first n terms is $\frac{a(r^n - 1)}{r - 1}$ or $\frac{a(1 - r^n)}{1 - r}$.

Glide reflection

The glide reflection is the single ***transformation*** that will replace the combined transformations: a ***translation parallel*** to ***line*** m and a ***reflection*** in line m, taken in either order.

Examples:

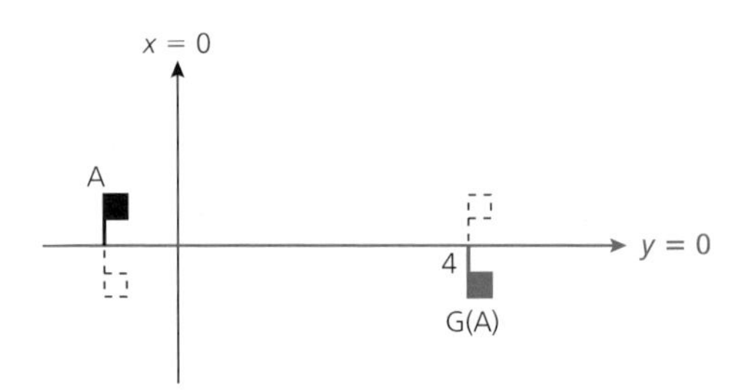

G(A) is the ***image*** after the glide reflection: reflect in $y = 0$ and translate $\begin{pmatrix} 5 \\ 0 \end{pmatrix}$

H(P) is the image after the glide reflection: translate parallel to m and reflect in line m.

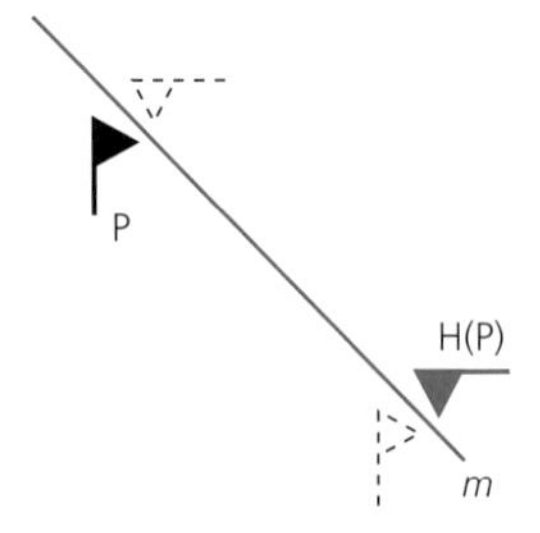

Gradient

The gradient of a ***line*** is a measurement of how 'steep' it is. The gradient of a line can be found by drawing ***right-angled triangles*** on the line.

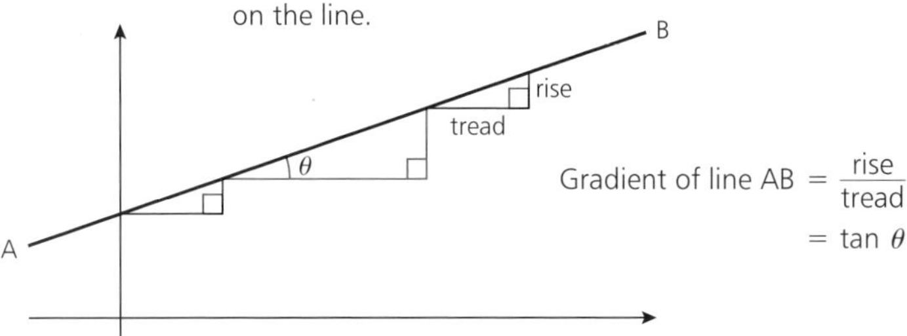

$$\text{Gradient of line AB} = \frac{\text{rise}}{\text{tread}} = \tan\theta$$

Example:

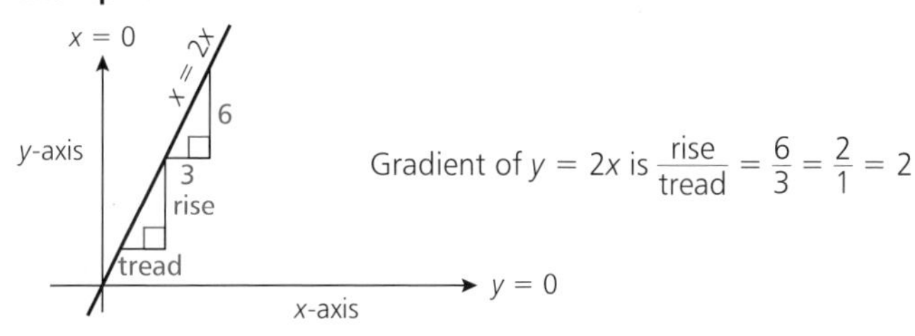

$$\text{Gradient of } y = 2x \text{ is } \frac{\text{rise}}{\text{tread}} = \frac{6}{3} = \frac{2}{1} = 2$$

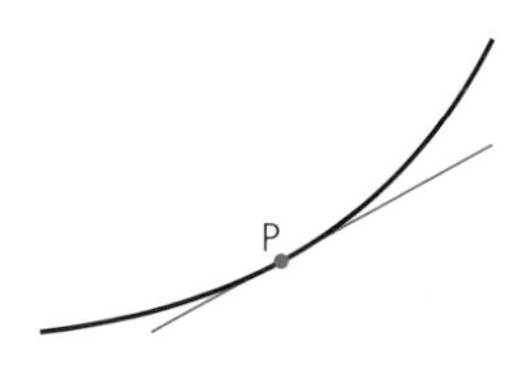

The gradient of a curve at a point on the curve is the gradient of the ***tangent to the curve*** at that point.

The gradient of the curve at point P can be estimated by drawing and found accurately from the value of the ***derivative*** of the ***function*** at that point.

Gram

A gram is a ***unit*** of ***mass***. A gram weight is the ***weight*** of a mass of one gram. (A large pea is about one gram weight.) The abbreviation for gram is g.

Graph

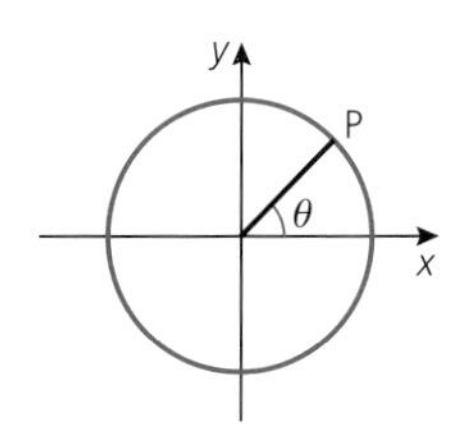

A graph is a way of illustrating information and numbers on paper to make it more easily understood.

Data are often displayed by ***bar charts***, ***pie charts***, ***frequency diagrams***, ***histograms***, and ***pictograms***.
In other situations ***axes*** and ***coordinates*** are used.

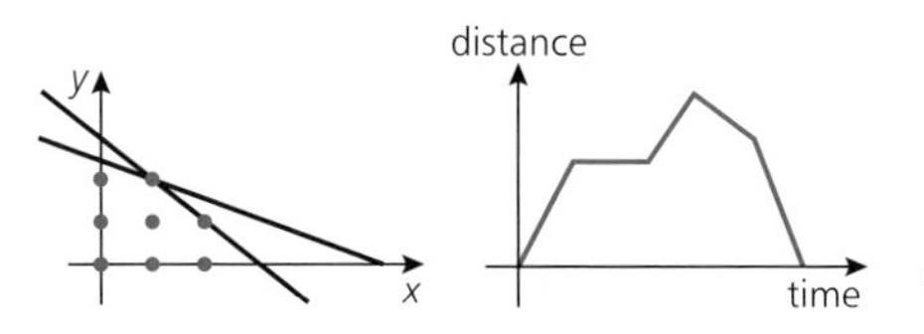

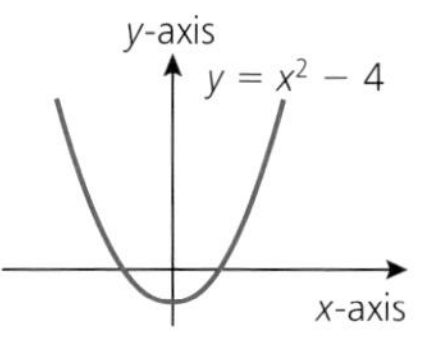

Great circle

A great circle is the largest ***circle*** that can be drawn on the surface of a ***sphere***.

The centre of a great circle is the centre of the sphere.

Greatest common divisor (GCD)

The greatest common divisor of a ***set*** of whole numbers is the largest ***whole number*** that divides into all of them without leaving a remainder.
The greatest common divisor is also called the ***highest common factor***.

H

HCF see **Highest common factor**

Hectare

A hectare (ha) is a measure of ***area*** and is usually used for areas of farms or other areas of land of similar size.

1 hectare = 10 000 square metres.

Hemisphere

A hemisphere is half a ***sphere***, formed by a ***plane of symmetry*** of the sphere.

Examples:

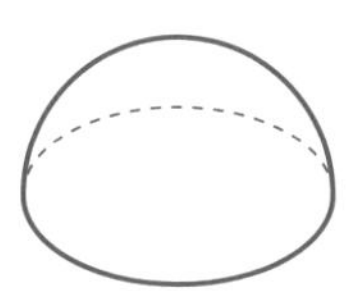

Hexagon

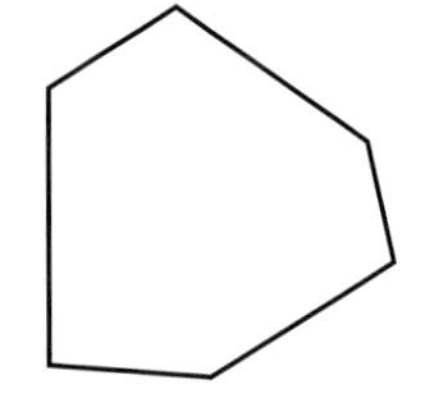

A hexagon is a ***polygon*** bounded by six straight ***lines*** and containing six ***angles***.

Examples:

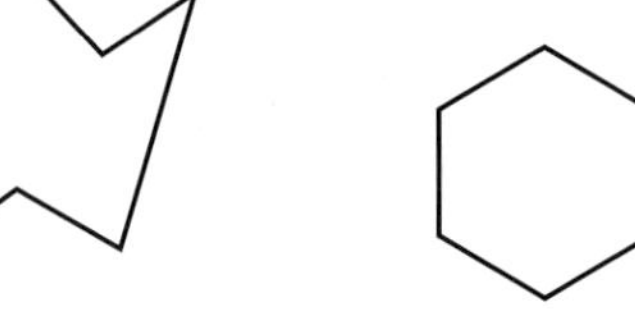

This is a ***regular*** hexagon.

Highest common factor (HCF)

Two ***counting numbers*** always have 1 as a common ***factor***. Often they will have other factors in common.

Example:
{factors of 12} = {1, 2, 3, 4, 6, 12}
{factors of 15} = {1, 3, 5, 15}
1 and 3 are the only common factors of 12 and 15.

The highest common factor (usually abbreviated to HCF) is the largest of these numbers, in this case 3.

Example:
{factors of 20} = {1, 2, 4, 5, 10, 20}
{factors of 30} = {1, 2, 3, 5, 6, 10, 15, 30}
10 is the highest common factor of 20 and 30.

Histogram

A histogram is similar to a ***bar chart***, but it is the ***area*** of the bars and not the height that shows the ***frequency***.

Example: The table shows the frequency f of families with x children in the family.

x	f
0	7
1	10
2	16
3	12
4	8
5	3
6 to 10	10

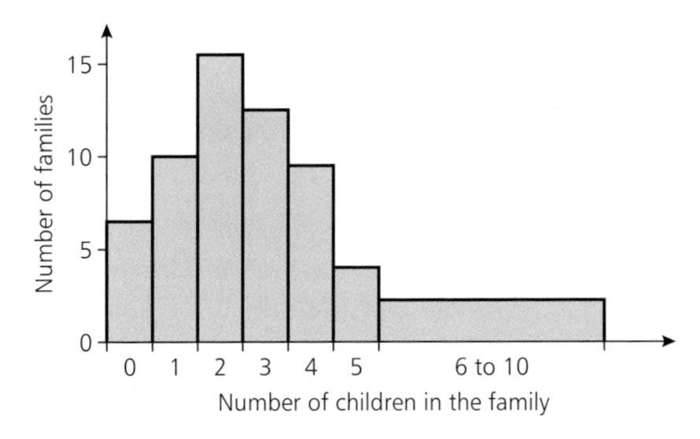

This is a histogram.

Horizontal

A line is horizontal if it is ***parallel*** to the Earth's skyline.

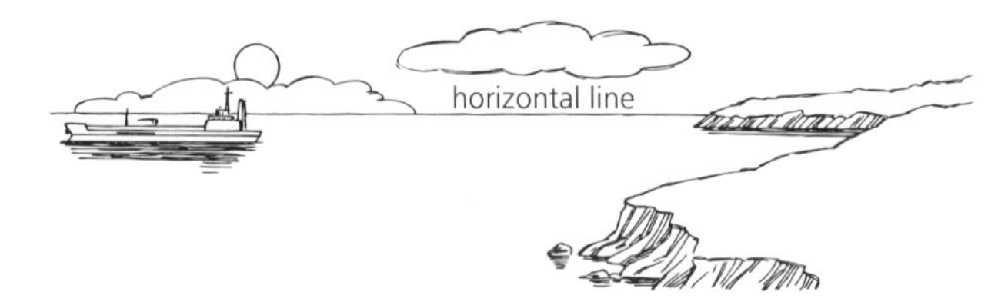

Example: When the water in a swimming pool is still, any line on its surface is horizontal.

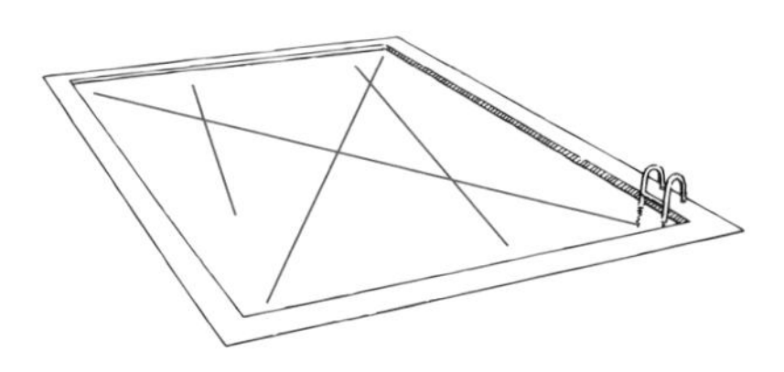

Hypotenuse

The hypotenuse is the side opposite the ***right-angle*** in a ***right-angled triangle***. It is also the longest side in a right-angled triangle.

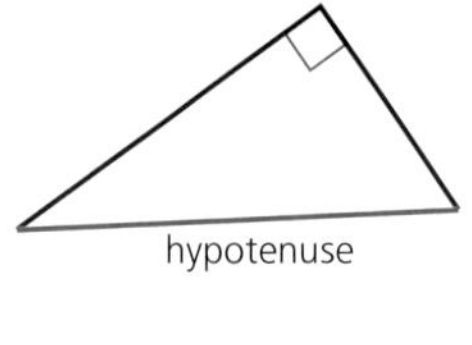

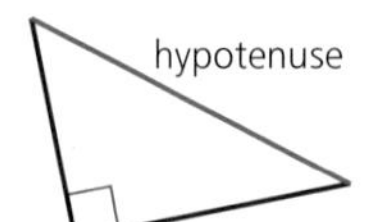

Examples:

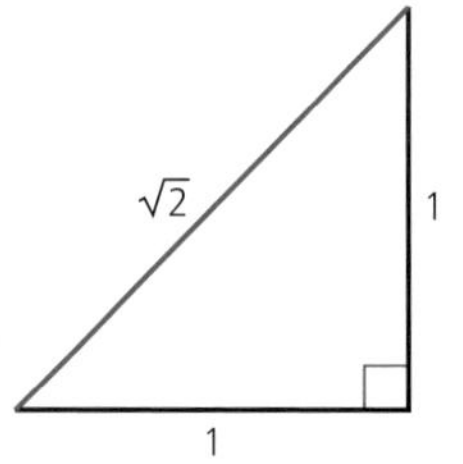

The hypotenuse is $\sqrt{2}$

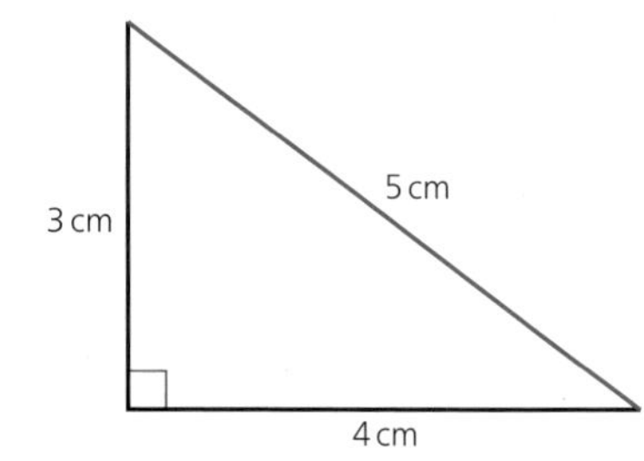

The hypotenuse is 5 cm

I

Icosahedron

An icosahedron is a solid with twenty ***faces***.

In a ***regular*** icosahedron each face is an ***equilateral triangle***.

Example (right):
This is a regular icosahedron.

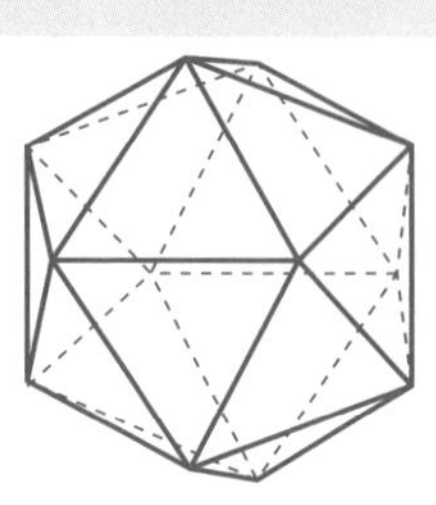

Identical

Shapes and objects are identical if they are exactly the same shape and size.
In some ***transformations***, such as ***reflection*** and ***rotation***, the ***object*** and ***image*** are identical.

Tessellations show patterns with one or more identical shapes, which fit together, without gaps, to fill space.
See **Congruent**

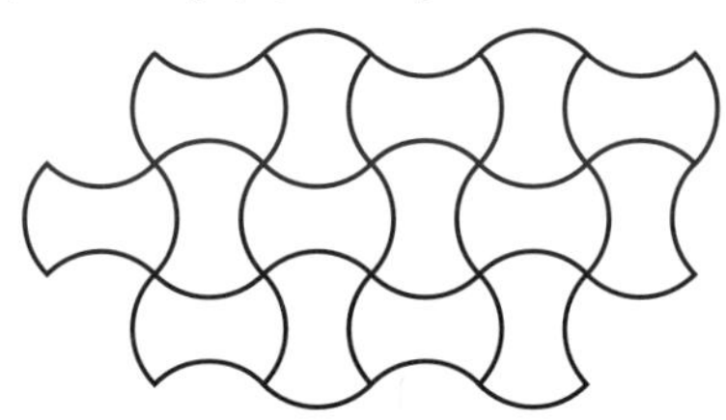

Identity

An identity is an ***equation*** where one side is a different way of writing the other side. The symbol $\equiv$ is sometimes used to show an identity. Identities are true (correct) for all values given to the letters.

Example: $a + b = b + a$ and $2(a + b) = 2a + 2b$ are identities.
See and contrast **Linear equation**

Identity element

An identity is an ***element*** which when combined with other elements leaves them the same. The identity depends on the ***set*** of elements and the ***operation***.

Examples:
For numbers

and the operation $\times$	and the operation $+$
it is 1	it is 0
$3 \times 1 = 3$	$0 + 3 = 3$
$2 \times 1 = 2$	$2 + 0 = 2$
$1 \times 5 = 5$	$0 + 5 = 5$
$-6 \times 1 = -6$	$-6 + 0 = -6$
$1 \times 4 = 4$	$9 + 0 = 9$

For sets

and the operation $\cup$	and the operation $\cap$
it is $\varnothing$	it is U
$A \cup \varnothing = A$	$P \cap U = P$
$\varnothing \cup B = B$	$U \cap Q = Q$

For 2 by 2 matrices

and the operation	and the operation
addition, it is $\begin{pmatrix} 0 & 0 \\ 0 & 0 \end{pmatrix}$	multiplication, it is $\begin{pmatrix} 1 & 0 \\ 0 & 1 \end{pmatrix}$

Identity mapping

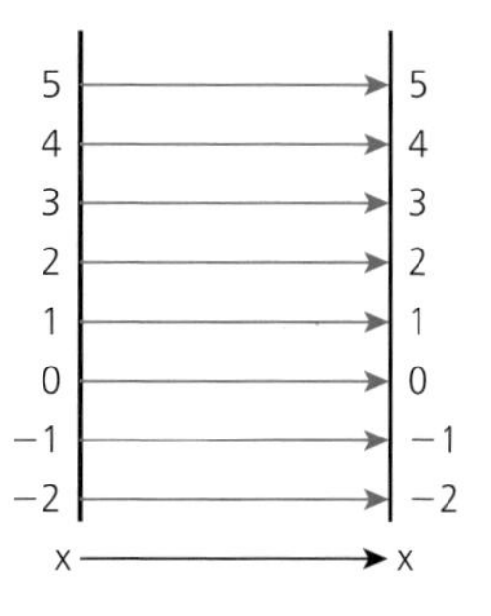

The identity mapping is the ***mapping*** that ***maps*** each ***element*** onto itself.

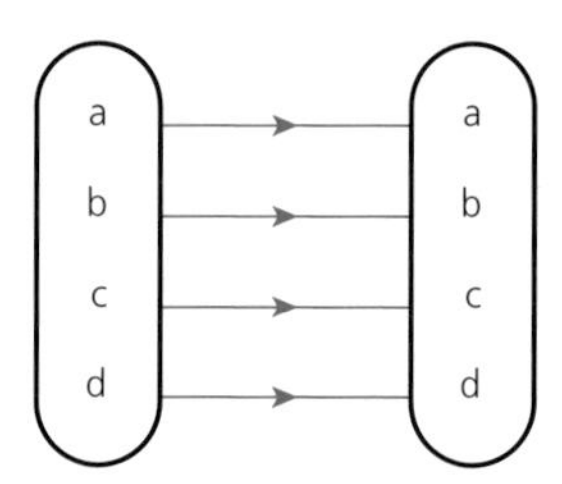

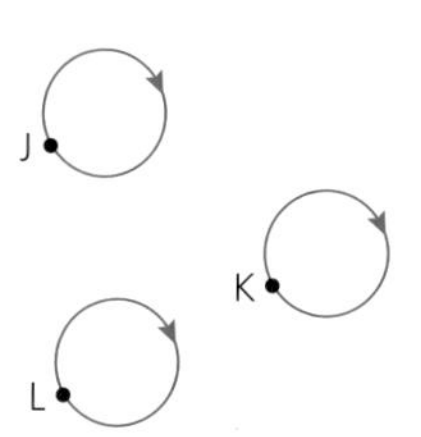

Identity transformation

The identity transformation is the ***transformation*** that ***maps*** each point of the ***plane*** onto itself. Under the identity transformation a shape remains unchanged and all properties of the shape are ***invariant***.

When ***matrices*** are used to represent transformations in the plane, the identity transformation is represented by the matrix $\begin{pmatrix} 1 & 0 \\ 0 & 1 \end{pmatrix}$.

Image

The image is the result when an ***object*** is ***transformed***.

Examples: The shape S′ is the image in each case.

Here the image is the result of a ***reflection***.

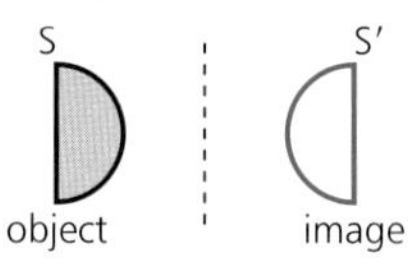

Here the image is the result of a ***rotation*** about 0.

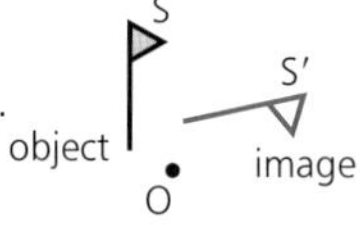

The image is also the result when a number undergoes a ***mapping***.

Examples:

$x \to x + 2$
$3 \to 5$
$2 \to 4$ the image of 3 is 5
$5 \to 7$ the image of 1 is 3
$1 \to 3$ the image of 6 is 8
$0 \to 2$
$6 \to 8$

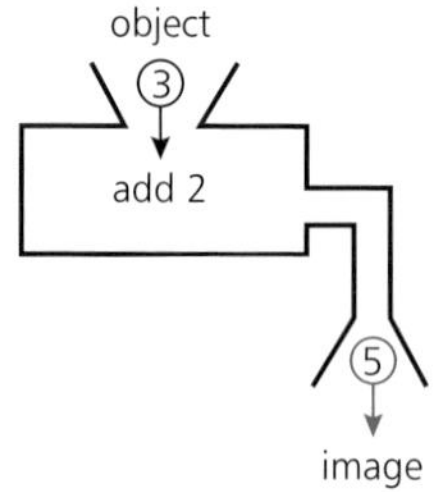

NOTE: The ***set*** of images {5, 4, 7, 3, 2, 8} is called the image set.

See **Corresponding points**

Improper fraction

An improper fraction is a ***fraction*** with the ***numerator*** bigger than the ***denominator***, such as

$\frac{5}{3}$, $\frac{7}{4}$, $\frac{10}{9}$, $\frac{16}{13}$, $\frac{8}{5}$, $\frac{11}{5}$

These improper fractions can be written as ***mixed numbers***.

Examples:

$\frac{4}{3} = 1\frac{1}{3}$, $\frac{8}{5} = 1\frac{3}{5}$

$\frac{11}{4} = 2\frac{3}{4}$, $\frac{15}{2} = 7\frac{1}{2}$

Indefinite integral

An indefinite integral, $\int f(x)dx$, is another ***function*** of x whose ***derivative*** is $f(x)$. Apart from a ***constant***, the other function is unique.

Example: $\int 2x dx = x^2 + c$ where c is a constant.

See **Definite integral**, **Integration**

Independent events

Two events are independent when the outcome of one event does not influence the outcome of the other event.

Example: When two coins are flipped, the way one coin lands has no effect on the way the other coin lands.

Index form

A number x written as the ***power*** of another number, a, is said to be in index form.
If $x = a^n$ then n is the index, showing the power of a.
The plural of index is indices.

Examples:

$49 = 7^2$
$16 = 2^4$
$10\,000 = 10^4$
$125 = 5^3$
$0.01 = 10^{-2}$

These numbers are written in index form; the indices are shown in green.

Inequality

An inequality is a statement showing the relationship between quantities that are not equal.

$a < b$ means 'a is less than b'

$a > b$ means 'a is greater than b'

$a \neq b$ means 'a is unequal to b'

Examples:

From the dots on the ***number line*** (*left*) it can be seen that

$4 > 1 \qquad 1 < 4$

$1 > -2 \qquad -2 < 1$

$-2 > -5 \qquad -5 < -2$

The ***solution*** to the inequality $x < 3$ is shown by the green line (*right*).

See **Directed numbers, Solution set**

Infinite set

A infinite set has an infinite, i.e. unlimited number of ***elements***.

Example: The ***set*** of ***whole numbers***, {0, 1, 2, 3, 4, ...} is an infinite set.

Inscribe

One geometric shape inscribes another when it is drawn inside and touches but does not cross the second shape.

Example: The ***circle*** inscribes the ***square*** and the ***triangle*** inscribes the circle.

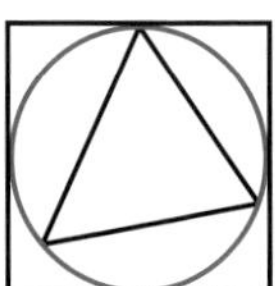

Integers

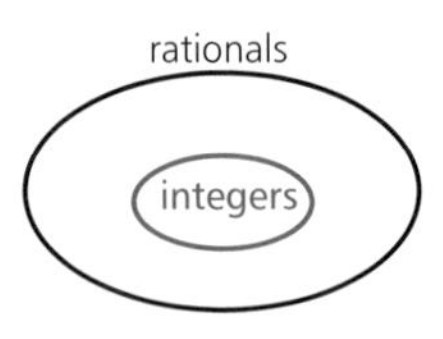

The ***set*** of integers is {... −4, −3, −2, −1, 0, 1, 2, 3, ...}

The set of integers is a ***subset*** of the set of the ***rationals***.

The set of ***natural numbers*** is a subset of the set of integers.

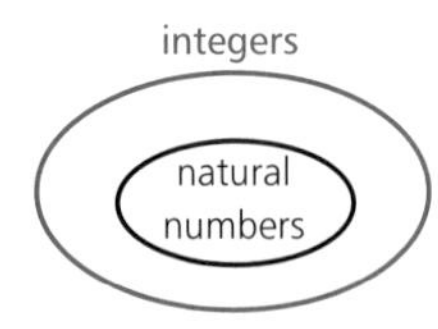

Integration

Integration is the process of finding the ***definite*** or ***indefinite integral*** of a ***function***.

Intercept

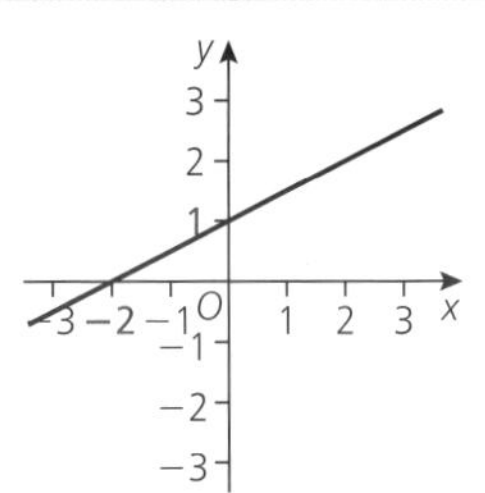

The intercept is the point where a ***line*** or curve crosses the *x*-***axis*** or the *y*-axis.

Sometimes it means the value of the ***coordinate*** at that point.

Example (left): The *y* intercept of this line is 1 and the *x* intercept is −2.

Interquartile range

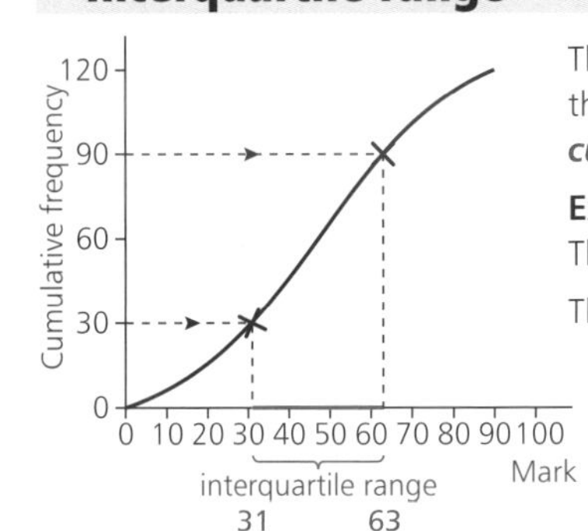

The interquartile range is the ***range*** of the ***data*** between the first and the third ***quartiles***. It is usually found from a ***cumulative frequency*** diagram.

Example (left):

The interquartile range is $63 - 31 = 32$ marks

The semi-interquartile range is $\frac{1}{2} \times (63 - 31) = 16$ marks.

Intersect

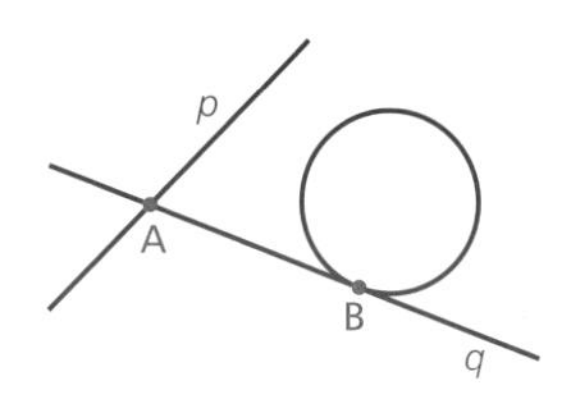

Lines and curves intersect if they cross or touch.

Examples:

Lines *p* and *q* (*left*) intersect at A;

the ***circle*** and line *q* intersect at B;

but the line *p* and the circle do not intersect.

Lines $y = 2x$ and $x + y = 3$ (*right*) intersect at the point (1, 2).

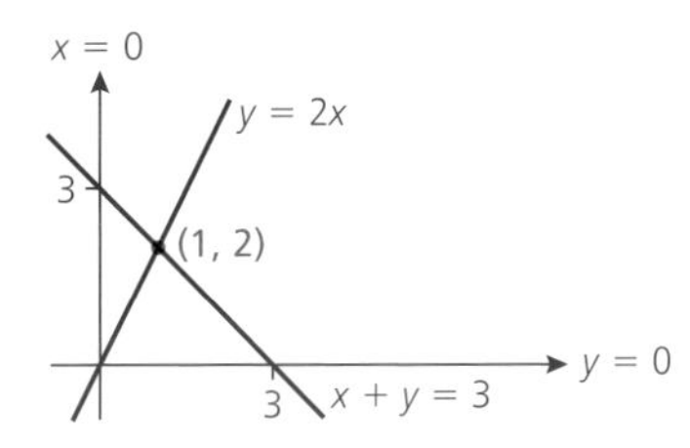

Point (2, 3) (*right*) is the intersection of line $x = 2$ and line $y = 3$.

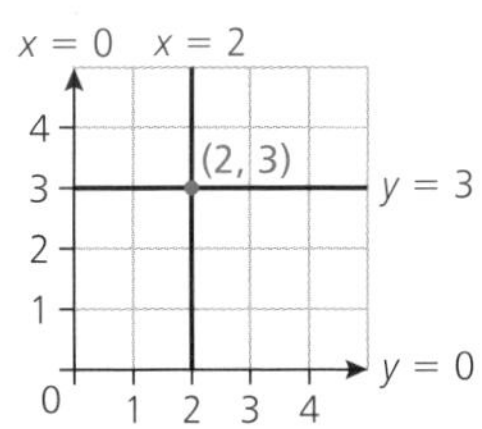

Intersection

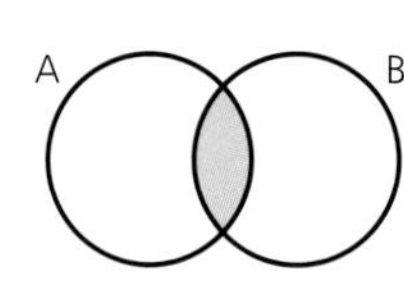

The intersection of two ***sets*** A and B is the set that has in it only those ***elements***, and all those elements, that are in both set A and set B.

Example:
If A = {2, 4, 6, 8}
and B = {1, 2, 4, 8}
then $A \cap B = \{2, 4, 8\}$
The intersection of sets is shown by the symbol $\cap$.

On a ***Venn diagram***, the intersection of set A and set B is shown by the set of points shaded green. $A \cap B$ is shaded green.

Invariants

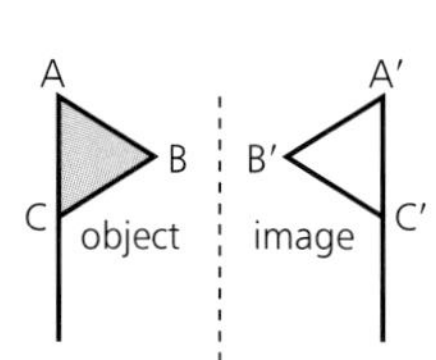

Invariants are facts that remain the same after a ***transformation***. Invariants are often properties of the ***object*** and ***image*** that are the same both BEFORE and AFTER the transformation, such as length, ***area***, ***angles***, ***parallel lines***.

Examples:
For a ***reflection***, some invariants are size of angle, area, and length of ***lines***.

Inverse element

The inverse of an ***element*** is the element which combines with it to give the ***identity***.
If a ***set*** has no ***identity element*** under an ***operation***, then there are NO inverses.

Examples:
For numbers under addition:

the inverse of 6 is −6 because $6 + -6 = 0$
the inverse of 3 is −3 because $3 + -3 = 0$
the inverse of −1 is 1 because $-1 + 1 = 0$
} and because 0 is the identity for numbers under addition.

For 2 by 2 ***matrices*** *under multiplication:*

the inverse of $\begin{pmatrix} 3 & 1 \\ 5 & 2 \end{pmatrix}$ is $\begin{pmatrix} 2 & -1 \\ -5 & 3 \end{pmatrix}$

because $\begin{pmatrix} 3 & 1 \\ 5 & 2 \end{pmatrix}\begin{pmatrix} 2 & -1 \\ -5 & 3 \end{pmatrix} = \begin{pmatrix} 1 & 0 \\ 0 & 1 \end{pmatrix}$

the inverse of $\begin{pmatrix} 1 & 0 \\ 7 & 6 \end{pmatrix}$ is $\frac{1}{6}\begin{pmatrix} 6 & 0 \\ -7 & 1 \end{pmatrix}$

because $\begin{pmatrix} 1 & 0 \\ 7 & 6 \end{pmatrix}\begin{pmatrix} 1 & 0 \\ \frac{-7}{6} & \frac{1}{6} \end{pmatrix} = \begin{pmatrix} 1 & 0 \\ 0 & 1 \end{pmatrix}$

and because $\begin{pmatrix} 1 & 0 \\ 0 & 1 \end{pmatrix}$ is the identity for 2 by 2 matrices and multiplication.

If $\mathbf{M} = \begin{pmatrix} a & b \\ c & d \end{pmatrix}$ then,

the inverse of **M** is $\mathbf{M}^{-1} = \frac{1}{\det \mathbf{M}} \begin{pmatrix} d & -c \\ -b & a \end{pmatrix}$ where

det **M** is the ***determinant*** of the matrix **M**.

Inverse function

For a ***function*** f that ***maps*** each point x onto its ***image***, y, the inverse mapping f^{-1} is the function that maps each point y of the ***image set*** back onto x.

Example: for $x \xrightarrow{f} 2x$ the inverse function is $x \xrightarrow{f^{-1}} \frac{1}{2}x$

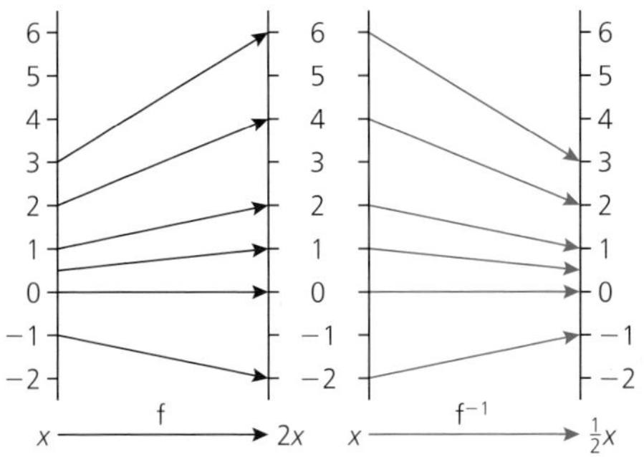

NOTE: A function f is either a MANY TO ONE or a ONE TO ONE ***correspondence***; BUT f can only have an inverse function if f is a ONE TO ONE correspondence.

Inverse proportion

Two quantities are inversely proportional when their ***product*** is constant.

Example: Distance = speed × time, so when a bus does a journey of 20 km, the relationship between its speed and the time taken is 20 = speed × time. The speed is inversely proportional to the time.
See **Direct proportion, Inverse variation, Joint variation, Proportion**

Inverse transformation

For a ***transformation*** **M** which ***maps*** point **P** onto its ***image*** **P′**, the inverse transformation $\mathbf{M}^{-1}$ is the transformation which maps **P′** back onto **P**, for all points.

Example: The transformation **E**: ***enlargement*** with ***scale factor*** 2 about O, has the inverse transformation $\mathbf{E}^{-1}$: enlargement with scale factor $\frac{1}{2}$ about O.

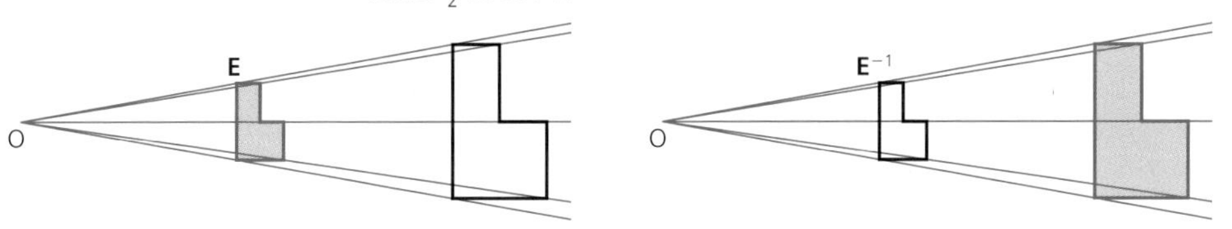

NOTE: If a transformation is represented by the ***matrix*** **A**, then the inverse transformation when it exists is represented by the ***inverse*** matrix $\mathbf{A}^{-1}$.

Inverse variation

Inverse variation is the same as ***inverse proportion*** but is usually used for two ***variables***.

If x and y are two variables, y varies inversely as x when $xy = k$ where k is a ***constant***.

This is usually written as $y = \frac{k}{x}$.

Irrational number

An irrational number is a ***number*** that cannot be written as a ***fraction*** $\frac{p}{q}$, where p and q are ***integers*** and q is not ***zero***.

See **Rational**

Examples:

$\sqrt{2} = 1.414\ldots$

$\pi = 3.14159265\ldots$

$\sqrt{10} = 3.1623\ldots$

$(\sqrt{7} + 8) = 10.6458\ldots$

When irrational numbers are expressed as ***decimals***, they do NOT ***terminate*** or ***recur***.

The ***union*** of the ***rational numbers*** and the irrational numbers is the ***set*** of ***real numbers***.

Isometric transformation

An isometric transformation does not alter the shape or size of an ***object***, just its position unless it is the ***identity transformation***. ***Rotations***, ***reflections*** and ***translations*** are isometric ***transformations***.

Isosceles

An isosceles ***triangle*** is one with two sides of equal length.

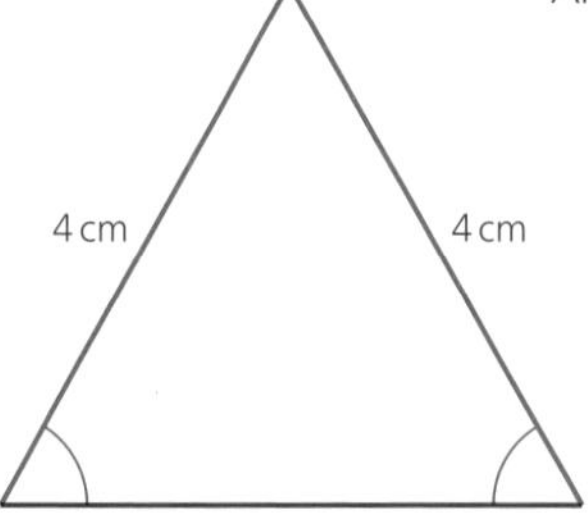

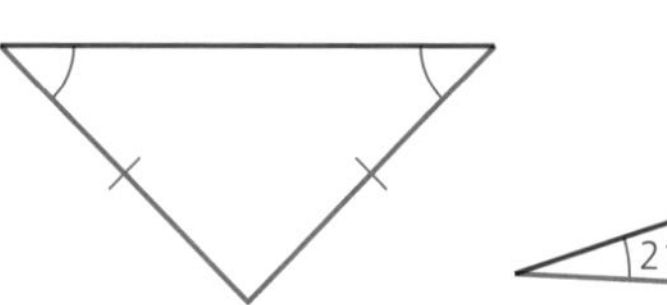

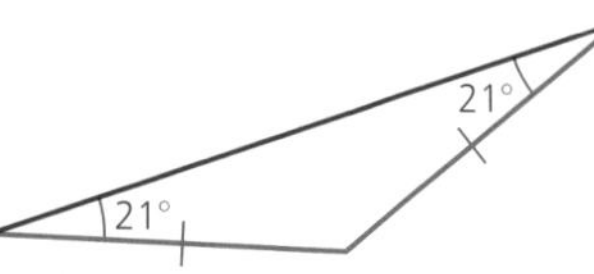

There are two equal ***angles*** in an isosceles triangle.

J

Joint variation

When a quantity is jointly ***variable*** to two or more other quantities it is ***directly proportional*** to each of them.

When z varies jointly with x and y, they are related by the ***equation*** $z = kxy$ where k is a ***constant***.

K

Kilogram

A kilogram is the standard ***unit*** of ***mass***. It is equal to 1000 ***grams***. We shorten kilogram to kg.

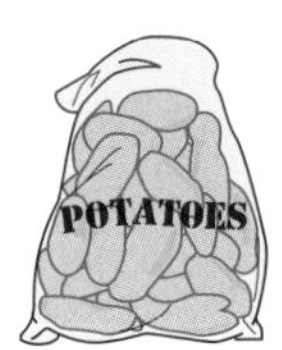

Examples:

A bag of potatoes may be 5 kg

A man may be 80 kg

1000 g = 1 kg
1000 kg = 1 ***tonne***

Kilometre

A kilometre is a length of 1000 ***metres***. We shorten kilometre to km. One kilometre is a distance of a little over $\frac{1}{2}$ a mile.

Kinematics

Kinematics is the study of the movement of objects using ***displacement***, ***velocity***, ***acceleration*** and time. Force and ***mass*** are not used.

Kite

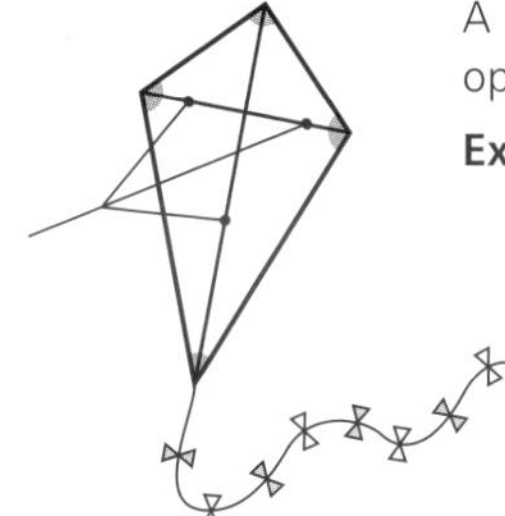

A kite is a ***quadrilateral*** with two pairs of equal sides which are not opposite to each other.

Examples:

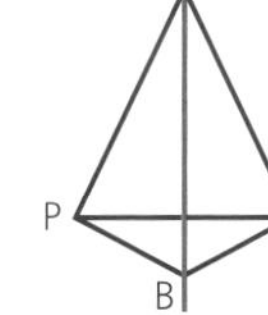

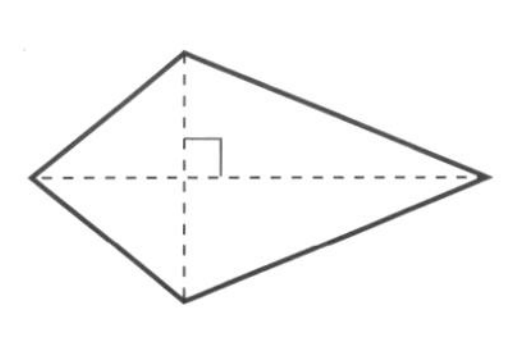

The ***diagonals*** of a kite cross at ***right-angles***.
A kite only has one ***line of symmetry***. AB is a line of symmetry but PQ is not.

Knot

A knot is a measure of ***speed***. It is 1 nautical mile per hour.
1 knot = 1.852 km/h

L

Latitude

A parallel of latitude is a ***circle*** on the surface of the Earth, with a centre on the line joining north and south poles. These circles are all ***parallel*** to each other and so are called parallels of latitude.

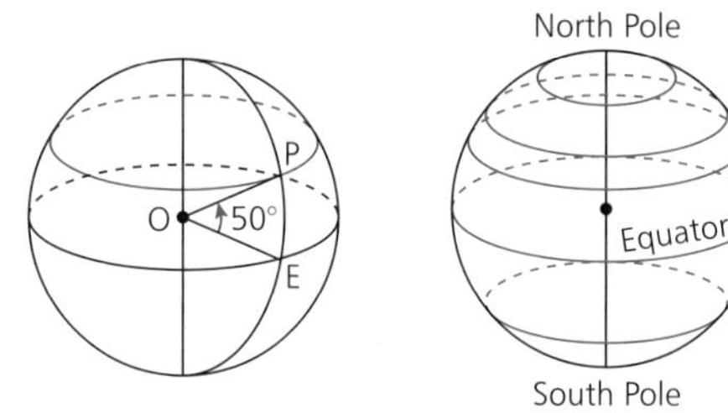

To describe any point P on the parallel of latitude shown in green we use the ***angle*** between the two ***radii*** of the Earth OP and OE. O is the centre of the earth and E is the point on the equator so that P and E are on the same line of ***longitude***. The green circle is called the parallel of latitude 50°N. Other parallels are similarly labelled.

LCM see **Least common multiple**

Leading diagonal

In a table or ***matrix*** which is ***square***, the line of numbers or letters going from the top left corner to the bottom right corner is called the leading diagonal.

Examples:

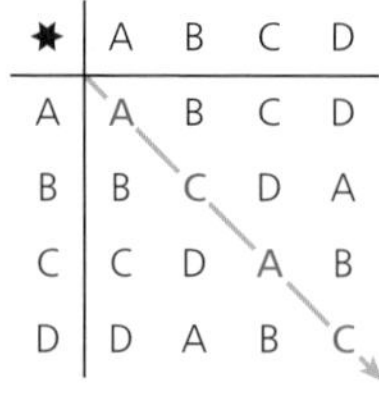

*	A	B	C	D
A	A	B	C	D
B	B	C	D	A
C	C	D	A	B
D	D	A	B	C

leading diagonal

$$\begin{pmatrix} 0 & 1 & 2 & 1 \\ 1 & 0 & 3 & 1 \\ 2 & 3 & 0 & 1 \\ 1 & 1 & 1 & 0 \end{pmatrix}$$

leading diagonal

Least common multiple (LCM)

For two (or more) numbers a and b, the common ***multiples*** are those numbers which appear in both the ***set*** of multiples of a and the set of multiples of b.

Example:

{multiples of 2} = {2, 4, 6, 8, 10, 12, 14, 16, 18, 20, 22, 24, ...}
{multiples of 3} = {3, 6, 9, 12, 15, 18, 21, 24, 27, ...}
{common multiples of 2 and 3} = {6, 12, 18, 24, ...}
The least common multiple (usually abbreviated to LCM) is the smallest of the common multiples.

Examples:

The LCM of 2 and 3 is 6
The LCM of 4 and 8 is 8
The LCM of 10 and 15 is 30
The LCM of 5 and 7 is 35

Line

A line can be straight or curved. It has no ends.

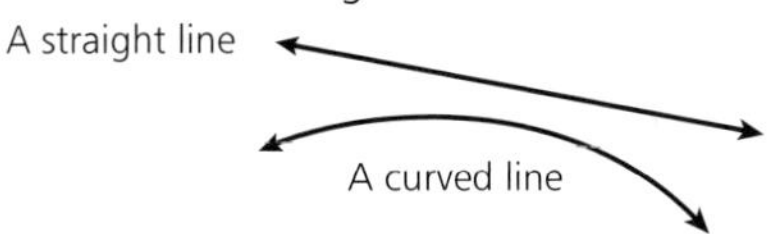

Line graph

A line graph is a ***graph*** where ***line segments*** are used to join points representing ***data*** such as temperatures at different times.

Example: This line graph shows the maximum temperatures in London for seven days in July 2006.

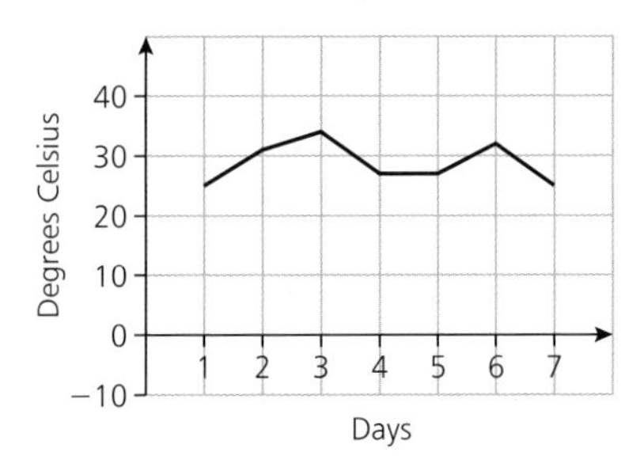

Line of best fit

A line of best fit is a ***line*** that goes approximately through the middle of a ***set*** of scattered points on a ***graph***.

Line segment

A line segment is a part of a straight ***line*** between two given points.

Example:

AB is a line segment

Line symmetry

A ***line*** of ***symmetry*** on a shape is a line which can be used as a fold, so that one half of the shape covers the other half exactly. A shape may have one or more lines of symmetry.

Examples:

one line of symmetry

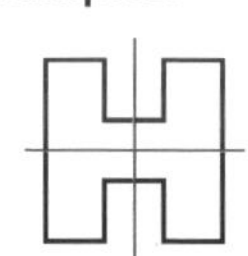

two lines of symmetry

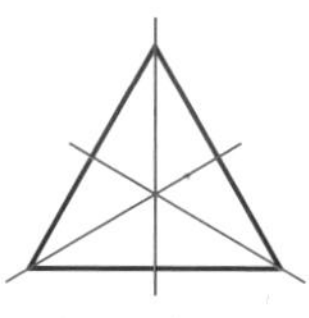

three lines of symmetry

no line of symmetry

Line symmetry is also called reflective symmetry because a mirror ***reflection*** on the line will produce the whole shape.
See **Mirror line**

Linear equation

A linear equation with ONE unknown (letter) is an ***equation*** which can be written in the form
$ax + b = 0$

Examples:

$3x + 4 = 7$ $\quad$ $2x + 6 = x - 2$

These equations can have only ONE ***solution***.

When a linear equation is used to show a ***linear function***, the equation can be written with the two unknowns x and y in the form

$$y = ax + b$$

In this case there are many solutions. See **Equation of a line**

Linear function

Linear functions are ***relations*** which can be written in the form $x \to ax + b$. When ***ordered pairs*** for a linear function are plotted on a ***graph*** they form a straight ***line***.

Examples: Here are some linear functions

$x \to x + 2$

$x \to 3x$

$x \to 4x - 1$

$x \to 5 - \frac{1}{2}x$

Linear inequality

A linear inequality in one unknown takes the form $ax + b > 0$ where a and b are ***constants***.

'>' means 'greater than'. Other signs that may be used are <, which means 'less than', $\leqslant$ which means 'less than or equal to' and $\geqslant$ which means 'greater than or equal to'.

The solution is a range of numbers that can be shown on a ***number line***.

A linear inequality in two unknowns takes the form $y < ax + b$.

The solution is the set of all the points in a ***region*** of the ***Cartesian plane***.

Examples: 1) The inequality $2x - 1 < 0$ can be solved as follows: $2x < 1$ so $x < \frac{1}{2}$

The open circle shows that $\frac{1}{2}$ is not included in the range.

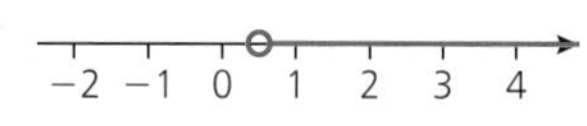

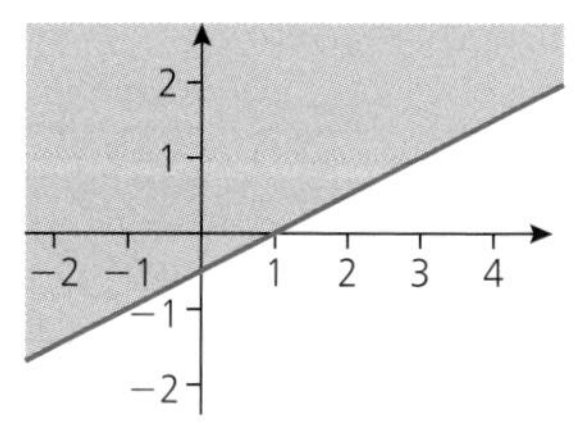

2) The solution of the inequality $y \geqslant \frac{1}{2}(x - 1)$ is the shaded region (*left*).

The solid line shows that the points on the line $y = \frac{1}{2}(x - 1)$ are included in the region.
See **Solution set**

Linear programming

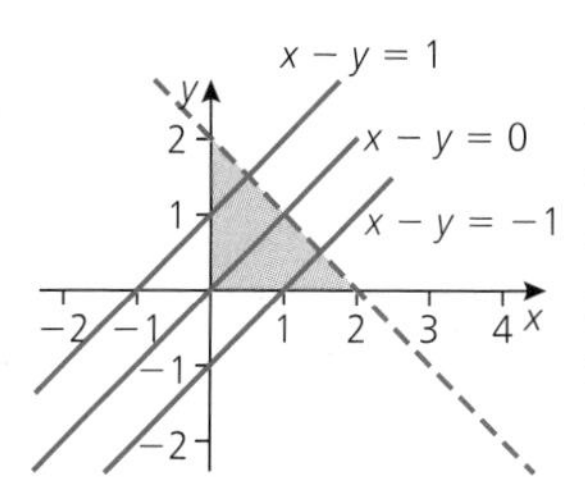

Linear programming is a method for finding the ***maximum*** or ***minimum value*** of a ***linear function*** when the unknowns are subject to constraints. The constraints are usually given as ***inequalities*** in x and y.

Example: The maximum ***integer*** value of $x - y$ is 1 when $x \geqslant 0$, $y \geqslant 0$ and $y > 2 - x$

Linear scale factor see **Scale factor**

Litre

A litre is a ***unit*** of ***volume*** equal to 1000 ***cubic centimetres***. We sometimes shorten litre to l, so long as there is no confusion with the number 1.

Locus (plural **Loci**)

The locus is the ***line*** or 'path' of a ***set*** of points that follow some rule or law.

Example:
If Q is a fixed point and {P:PQ is 2 cm} then the locus of P is a ***circle***.

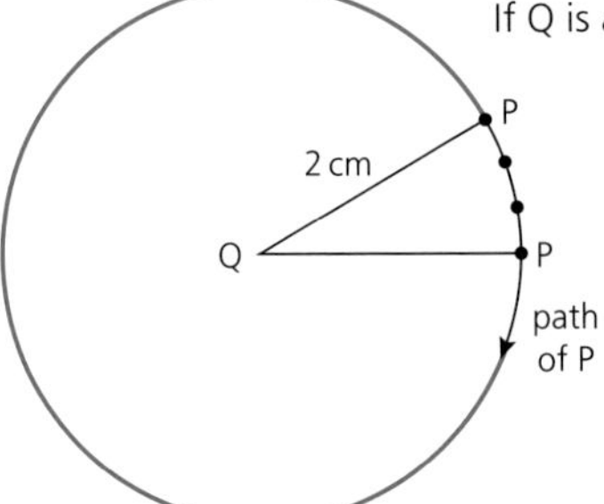

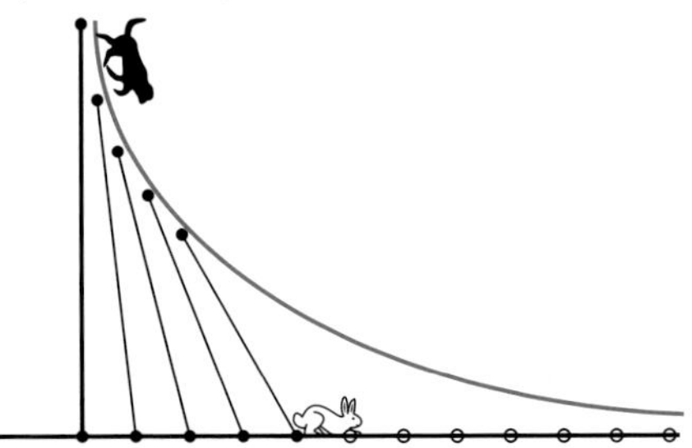

The locus of the dog chasing the rabbit is shown in green.

Logarithm

A logarithm is an ***index***. The base of the logarithm is the number raised to a ***power***, so the logarithm of a number is the power to which the base is raised to give that number.
$\log_2 8 = 3$ means 3 is the power to which 2 is raised to give 8, so $\log_2 8 = 3$ means $2^3 = 8$.

Common logarithms have a base of 10. Tables of common logarithms of all numbers between 1 and 10 are used in computations.

The **laws of logarithms** are $\log(a \times b) = \log a + \log b$
$\log(a \div b) = \log a - \log b$
$\log a^n = n \log a$

Longitude

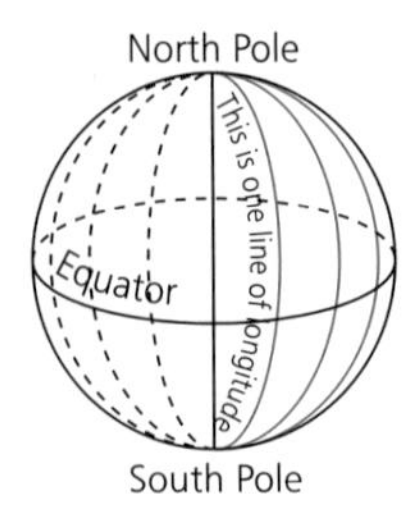

A line of longitude is half a ***circle*** on the surface of the Earth with one end at the South Pole and one end at the North Pole.

To describe any point P on the line of longitude shown in green we use the ***angle*** between the ***radii*** OG and OH. O is the centre of the Earth, G and H are points on the equator, and H is also on the green line of longitude. G is always on the black line of longitude called the Greenwich meridian. It is a fixed line of longitude passing through Greenwich, London in the UK. The green line is the line of longitude 80°E. Other lines of longitude are similarly labelled.

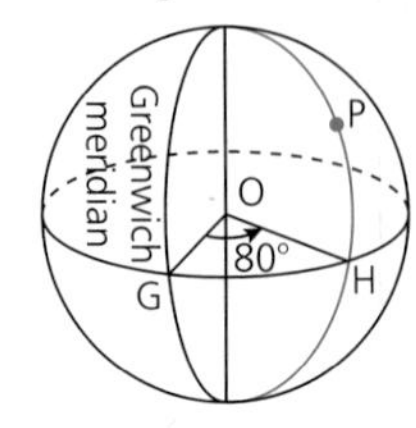

Loss

A loss is made when the selling price of an object is less than the purchase price.

M

Magnitude of a vector

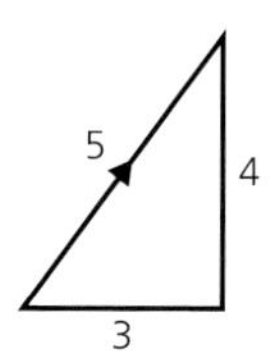

The magnitude of a ***vector*** is the length of the ***line*** that represents it.

Example: The magnitude of the vector $\begin{pmatrix} 3 \\ 4 \end{pmatrix}$ is 5.

Mantissa

The mantissa is the decimal part of a common ***logarithm***. The numbers in the body of tables of common logarithms are mantissas.

Example: $\log_{10}2 = 0.3010$

$\log_{10}200 = 2.3010$

Mapping

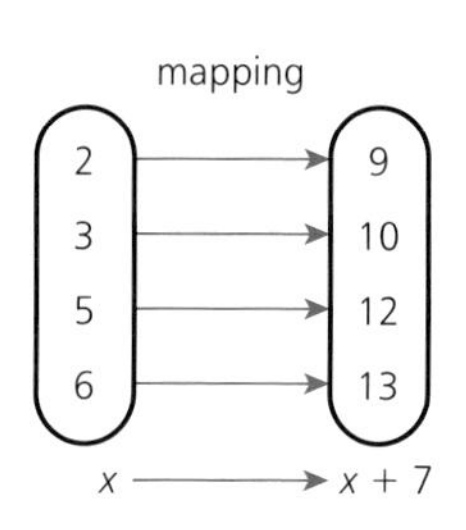

A mapping is a ***relation*** in which each ***object*** maps to only ONE ***image***. A mapping is the same as a ***function***.

Examples (left and right):

See **Correspondence**, **Function**

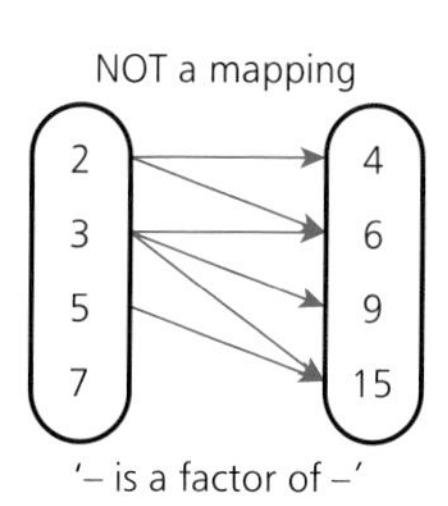

Mapping diagram

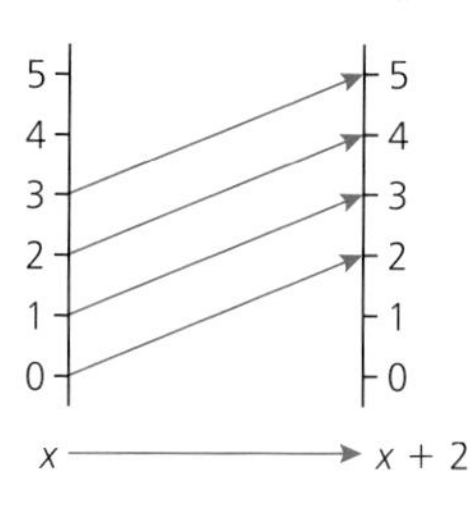

A mapping diagram is a diagram to show a ***mapping***.

Example (left):

Mean

(More correctly its full name is Arithmetic mean)

The mean is often just called the ***average*** by most non-mathematicians.

The mean of a ***set*** of n numbers is the ***sum*** of the n numbers ***divided*** by n.

Example: The mean of 1, 4, 3, 0, 1, 2, 1, 4 (eight numbers)
is $(1 + 4 + 3 + 0 + 1 + 2 + 1 + 4) \div 8$
$= 16 \div 8$
$= 2$
2 is the mean.

Mean deviation

The mean deviation is a measure of the spread of ***data***. It is calculated by finding the sum of the ***differences*** between every value and the ***mean*** of the values and then ***dividing*** this result by the number of values.

The ***formula*** for the mean deviation is $\frac{1}{n}\sum_{i=1}^{n} |\bar{x} - x_i|$ where $|\bar{x} - x_i|$ means the positive value of (the mean − a value).

Median

The median of a ***set*** of numbers is the middle number once the numbers have been arranged in order of size.

Examples:

a) The median of 2, 0, 3, 1, 4, 1, 5, 3, 2
is found from 0, 1, 1, 2, 2, 3, 3, 4, 5
2 is the median

b) The median of 1, 4, 3, 0, 1, 2, 1, 4
is found from 0, 1, 1, 1, 2, 3, 4, 4

There are two middle values so $1\frac{1}{2}$ is the median

NOTE: For a large collection of ***data***, the median is the value for which half the data is greater and half the data is less. In this case, the median is usually found from a ***cumulative frequency*** curve.

See **Quartiles**

Median of a triangle

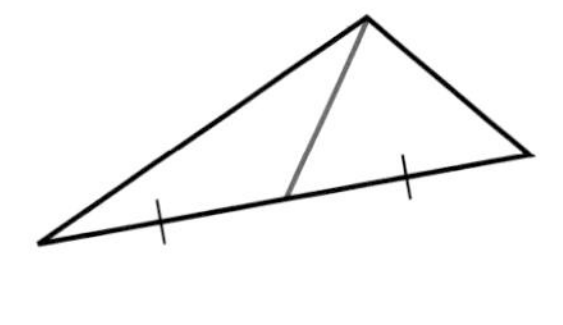

A median of a triangle is the line drawn from a ***vertex*** of the ***triangle*** to the mid-point of the side opposite the vertex.

Examples:

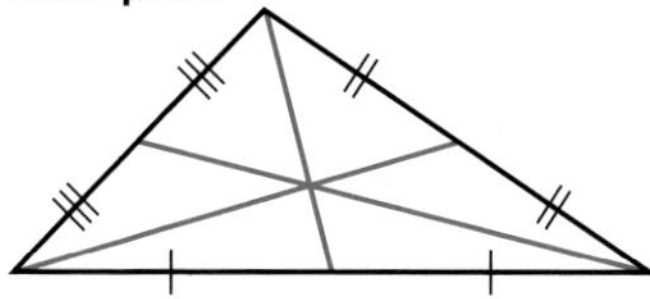

There are three medians for every triangle, one from each vertex, and all of them pass through the same point.
See **Centroid**

Mediator

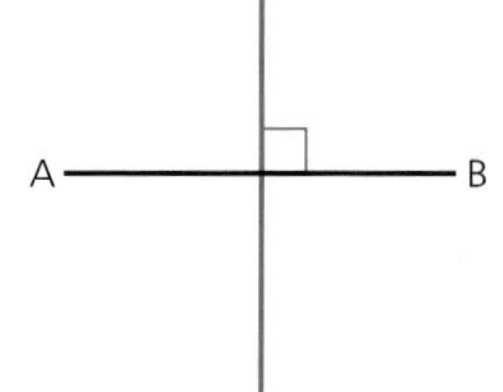

The mediator of a ***line segment*** AB is the ***line*** which cuts AB in half at ***right-angles***.
The mediator is also the ***line of symmetry*** for the line segment.

Examples:

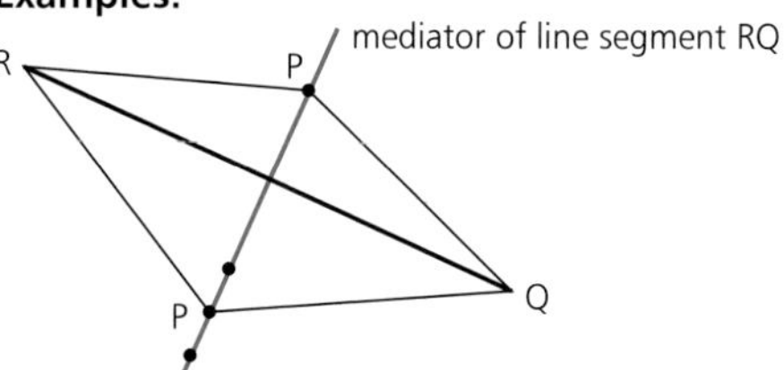

NOTE: The ***locus*** of points P which are equidistant from R and Q is the mediator of RQ.

See **Bisector**, **Perpendicular bisector**

Member

The ***elements*** of a ***set*** are the members of the set.

The symbol for 'is a member of' is ∈

Examples:
Dog is a member of the set of animals
Dog ∈ {animals}

5 is a member of the set of counting numbers.
$5 \in \{1, 2, 3, 4, 5, 6, 7, 8, \ldots\}$

Meridian

A meridian is a ***great circle*** on the surface of the Earth that goes through the north and south poles. A meridian is also called a circle of ***longitude***.

Metre

The metre is the standard ***unit*** of length. We shorten metre to m.

Example: A door is about 2 metres high.

Mid-ordinate rule

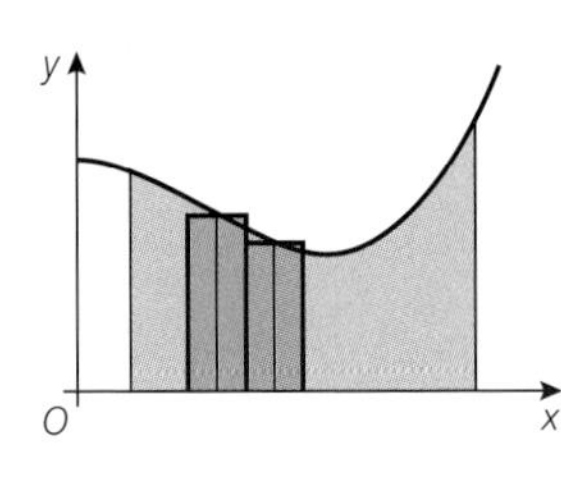

The mid-ordinate rule is a method for ***estimating*** the ***area*** enclosed by a curve, the *x*-***axis*** and two values of *x*.

The area is divided into equal width strips by ***lines parallel*** to the *y*-axis.

The area of each strip is estimated as the area of the ***rectangle*** whose height is the ***ordinate*** midway along the width.

The total area is approximately the sum of the areas of these rectangles.

Milligram

A milligram is a ***unit*** of ***mass***. 1 milligram = $\frac{1}{1000}$ ***gram***.
We shorten milligram to mg.

Millimetre

A millimetre is one thousandth of a ***metre***. We shorten millimetre to mm.

1 mm = $\frac{1}{1000}$ m

This gap is one millimetre

Minimum value

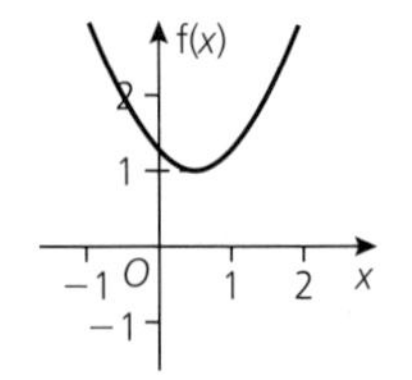

A minimum value of a ***function*** is its least value in a ***region*** around that value.

Example:
The minimum value of this function is 1.

Mirror line

The mirror line is the name given to the ***line*** in which an ***object*** is ***reflected***.

Mixed number

A mixed number is a number which is a ***whole number*** with a ***fraction***.

Examples:
$3\frac{1}{2}$, $2\frac{1}{5}$, $7\frac{3}{10}$, $4\frac{3}{5}$, $18\frac{3}{4}$
NOTE: $3\frac{1}{2}$ is the ***sum*** $3 + \frac{1}{2}$
See **Improper fraction**

Mode

The mode is the 'most popular' value, or the most frequently occurring item.

Example:
For 6, 7, 7, 8, 6, 5, 9, 8, 7, 6,
5, 9, 8, 7, 4, 7, 8, 6, 8, 7,
8, 5, 7, 7, 3, 7, 7, 3,

There are two 3s
There is one 4
There are three 5s
There are four 6s
There are ten 7s   7 comes TEN times
There are six 8s
There are two 9s

7 is the mode, it comes most often.

NOTE: If there are TEN 7s and TEN 8s, then both 7 and 8 are modes.

Modular arithmetic

Modular arithmetic uses the ordinary rules of arithmetic with ***whole numbers*** except that answers are replaced by the remainder when they are ***divided*** by the modulus.

Example: When the modulus is 6, $5 \times 4 = 2$ because $20 \div 6 = 3$, remainder 2 .

Multiple

A multiple of a number n is $k \times n$ where k is a ***natural number***.

Examples:
Some multiples of 4 are 8, 12, 20, 24, 28
Some multiples of 5 are 5, 15, 20, 25, 35
{multiples of 3} = {3, 6, 9, 12, 15, 18, 21, 24, ...}
{multiples of 7} = {7, 14, 21, 28, 35, 42, ...}

Multiplicative inverse

A multiplicative inverse is an ***inverse*** under multiplication.

Examples:

For numbers:

The multiplicative inverse of 3 is $\frac{1}{3}$ because $3 \times \frac{1}{3} = 1$

The multiplicative inverse of $\frac{1}{2}$ is 2 because $\frac{1}{2} \times 2 = 1$

The multiplicative inverse of -4 is $\frac{-1}{4}$ because $-4 \times \frac{-1}{4} = 1$

and 1 is the ***identity element*** for numbers under multiplication.

For 2 by 2 matrices:

The multiplicative inverse of $\begin{pmatrix} 2 & 7 \\ 1 & 4 \end{pmatrix}$ is $\begin{pmatrix} 4 & -7 \\ -1 & 2 \end{pmatrix}$

because $\begin{pmatrix} 2 & 7 \\ 1 & 4 \end{pmatrix}\begin{pmatrix} 4 & -7 \\ -1 & 2 \end{pmatrix} = \begin{pmatrix} 1 & 0 \\ 0 & 1 \end{pmatrix}$ and $\begin{pmatrix} 1 & 0 \\ 0 & 1 \end{pmatrix}$

is the identity for 2 by 2 matrices under multiplication.

See **Inverse elements**

Mutually exclusive

Mutually exclusive events cannot both happen.

Example: When a coin is flipped, getting a head and getting a tail are mutually exclusive because it is possible to get a head OR a tail. It is not possible to get a head AND a tail.

N

Natural numbers

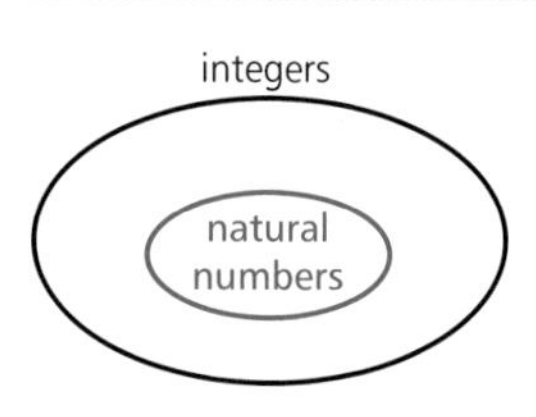

The ***set*** of natural numbers is another name for the set of ***counting numbers***.

The set of natural numbers is {1, 2, 3, 4, 5, 6, 7, ...}

The set of natural numbers is usually represented by the symbol ℕ, and is a ***subset*** of the set of ***integers***.

See **Whole numbers**

Negative number

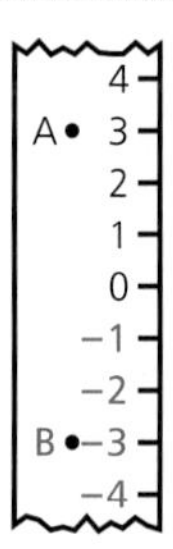

Negative numbers are used to count or measure in the opposite sense to the ***positive numbers***. They are marked with a − sign.

{negative ***integers***} = {...−8, −7, −6, −5, −4, −3, −2, −1}

See **Directed numbers**

Example (left): Point A is 3 above ***zero*** and is a positive number. Point B is 3 below zero and is a negative number, −3.

NOTE: Negative numbers enable us to find an ***additive inverse*** for all ***real numbers***.

Net

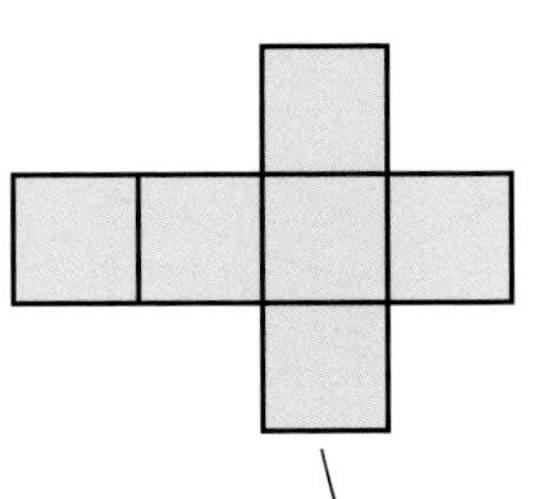

The net of a solid is the ***plane shape***, which when cut out and folded, can be made into the solid shape.

Example: The shape shown in green can be cut out and made into the ***cube***. It is one possible net for a cube.

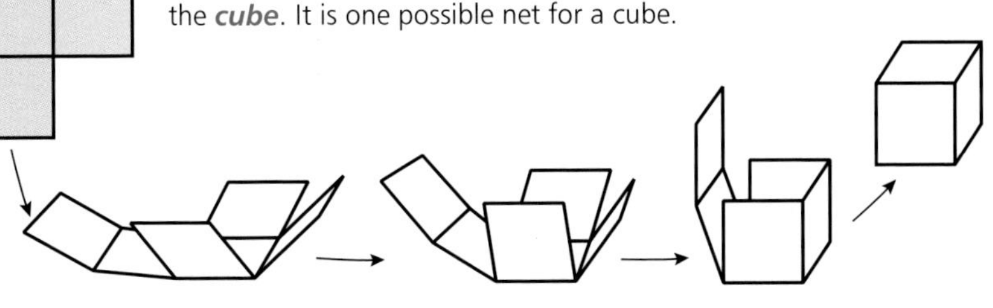

Nought

Nought is the symbol, 0, that stands for ***zero***.

Null set

The null set is another name for the ***empty set***. It is the ***set*** which has NO ***members*** and the symbol for it is ∅, or { }.

Number line

A number line shows the ***real numbers*** as points on a line.

$-4\ -3\ -2\ -1\ 0\ 1\ 2\ 3\ 4$

Numerator see Fraction

O

Object

In a ***transformation*** the shape or point being transformed is called the object.

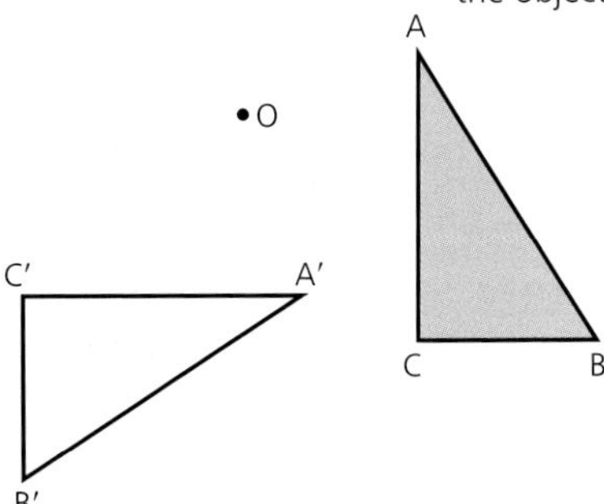

Example:
If the ***triangle*** ABC is ***rotated*** $-90°$ about O, the result is triangle A′B′C′.

The shape ABC, shown in green, is called the object; and the point A is an object point, for which A′ is its ***image***.

See **Image**

In a ***mapping*** any number in the ***set*** being mapped is an object, but the whole set being mapped is usually called the ***domain***.

Obtuse angle

An obtuse angle is an ***angle*** more than 90° but less than 180°.

Examples:

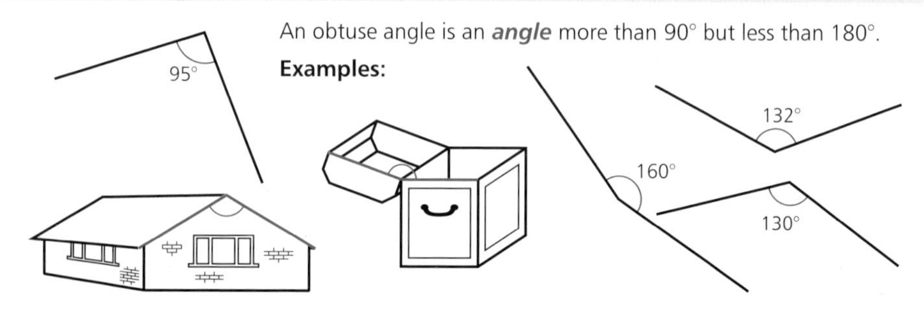

Octagon

An octagon is a ***polygon*** bounded by eight straight ***lines*** and containing eight ***angles***.

Examples:

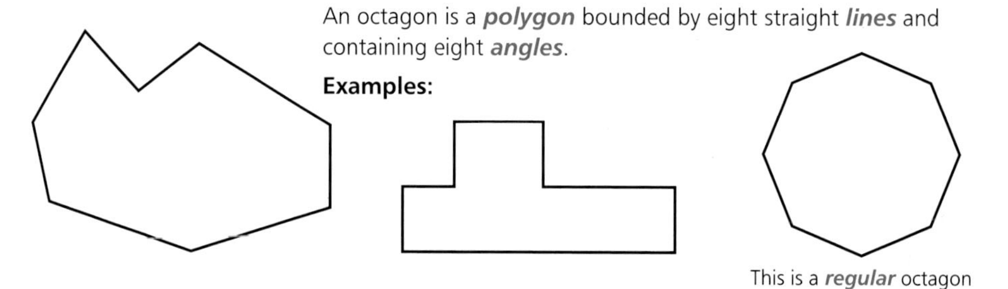

This is a ***regular*** octagon

Octahedron

An octahedron is a solid shape with eight ***faces***. In a ***regular*** octahedron each face is an ***equilateral triangle***.

Example:

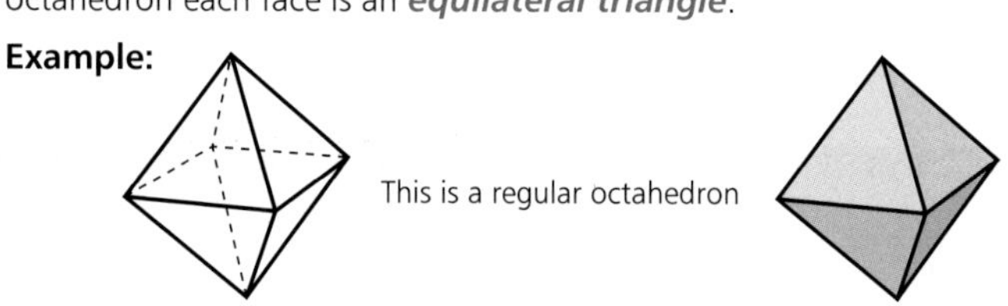

This is a regular octahedron

Odd numbers

Odd numbers are ***natural numbers*** that are not ***even***. Odd numbers do not divide exactly by 2. Odd numbers are those ending in 1 or 3 or 5 or 7 or 9.

Examples: These are odd numbers:
7, 13, 21, 39, 403, 60 827, 74 155, 20 0461.

Ogive

An ogive is another name given to a ***cumulative frequency*** curve.

Operation

An operation is a way of combining ***elements*** in a ***set***. The commonest kind of operation is a binary operation (binary meaning two). A binary operation combines two elements into one element.

Examples:

$\times$ is a binary operation on numbers

$4 \times 2 = 8 \qquad \frac{1}{2} \times 6 = 3 \qquad 1 \times 9 = 9 \qquad 3 \times \frac{1}{3} = 1$

$+$ is a binary operation on numbers

$2 + 3 = 5 \qquad 1 + 7 = 8 \qquad 0 + 9 = 9 \qquad -3 + 3 = 0$

$\cup$ is a binary operation on sets

$A \cup B = C \qquad A \cup C = D \qquad A \cup \varnothing = A$

$+$ is a binary operation on vectors

$\begin{pmatrix} 2 \\ 3 \end{pmatrix} + \begin{pmatrix} 5 \\ 1 \end{pmatrix} = \begin{pmatrix} 7 \\ 4 \end{pmatrix} \qquad \begin{pmatrix} 0 \\ 1 \end{pmatrix} + \begin{pmatrix} 5 \\ -6 \end{pmatrix} = \begin{pmatrix} 5 \\ -5 \end{pmatrix} \qquad \begin{pmatrix} 0 \\ 0 \end{pmatrix} + \begin{pmatrix} 7 \\ 9 \end{pmatrix} = \begin{pmatrix} 7 \\ 9 \end{pmatrix}$

multiplication is a binary operation on ***matrices***

$\begin{pmatrix} 1 & 1 \\ 2 & 3 \end{pmatrix} \begin{pmatrix} 4 & 1 \\ 0 & 2 \end{pmatrix} = \begin{pmatrix} 4 & 3 \\ 8 & 8 \end{pmatrix} \qquad \begin{pmatrix} 1 & 0 \\ 0 & 1 \end{pmatrix} \begin{pmatrix} 2 & 1 \\ 3 & 4 \end{pmatrix} = \begin{pmatrix} 2 & 1 \\ 3 & 4 \end{pmatrix}$

See **Structure**

Order of a matrix

The order of a matrix is the size of the ***matrix***. If a matrix has m rows and n columns, then the matrix is of order m by n.

Examples:

A matrix of order 2 by 3 is this size $\begin{pmatrix} \square & \square & \square \\ \square & \square & \square \end{pmatrix}$

If $\mathbf{A} = \begin{pmatrix} 1 & 0 & 5 & 1 \\ 3 & 1 & 6 & 1 \\ 9 & 0 & 0 & 1 \end{pmatrix}$, then **A** is a 3 by 4 matrix.

If $\mathbf{B} = \begin{pmatrix} 7 & 8 & 1 & 5 & 0 \end{pmatrix}$, then **B** is a 1 by 5 matrix.

Order of rotational symmetry see **Rotational symmetry**

Ordered pairs

When we use pairs of numbers and the order is important they are called ordered pairs.

Examples:

Coordinates are ordered pairs: (2, 3) is not the same point as (3, 2). The order 2, 3 matters.

The ordered pair (1, 4) is a ***solution*** to the ***equation***
$y = x + 3$ because $\boxed{4} = \boxed{1} + 3$ but (4, 1) is not a solution
$\boxed{1} \neq \boxed{4} + 3$; the order 1, 4 matters.

Ordinate

An ordinate is the *y-**coordinate*** of a point on the ***Cartesian plane***.

Origin

The origin is the point with ***coordinates*** (0, 0).
It is the point where the ***axes*** cross.
It is the point where the ***line*** $x = 0$ and the line $y = 0$ ***intersect***.
The origin is the point shown by the green dot.

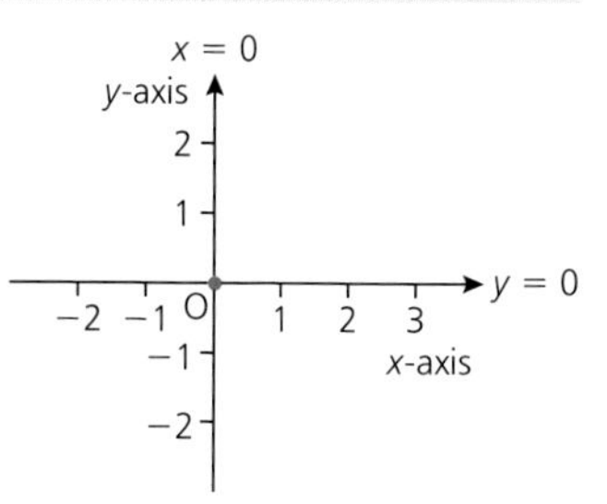

Orthocentre

The orthocentre is the point of ***intersection*** of the ***altitudes of a triangle***.
See and compare with **Centroid**

P

Parabola

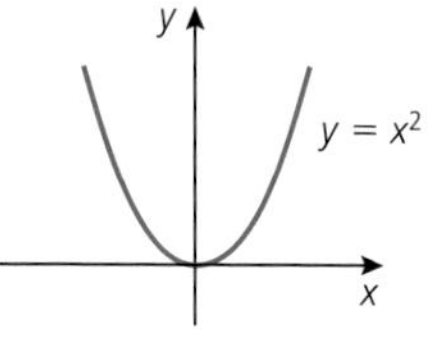

This is the shape obtained by plotting the ***function*** $x \rightarrow x^2$ using ***coordinates***. It is also the shape for any function of the form $x \rightarrow ax^2 + bx + c$.

The shape is approximately the path followed by a ball in flight, or the cross-section of the reflector for an electric bar fire.

Examples:

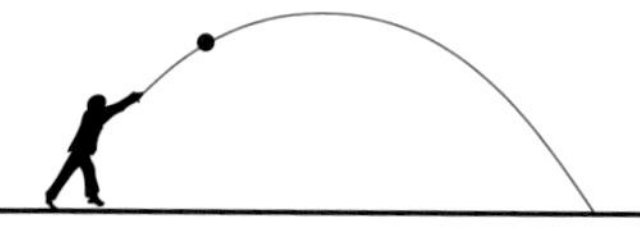

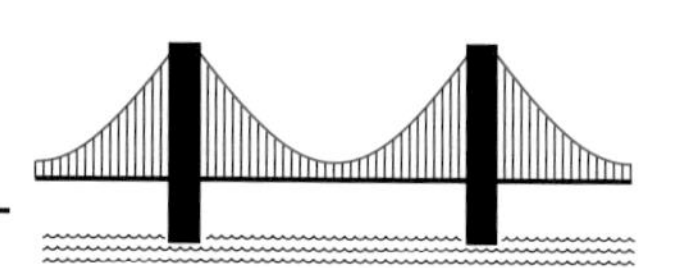

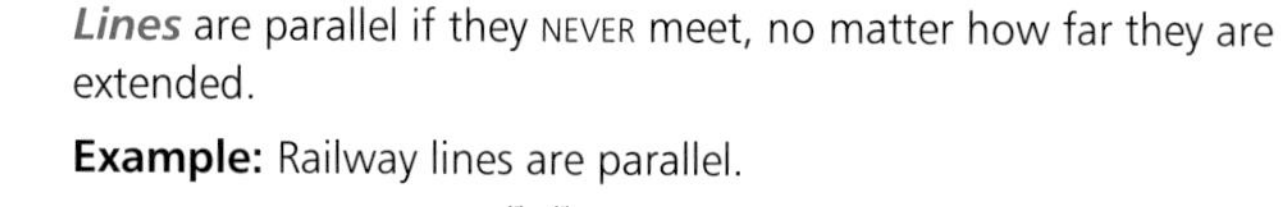

Parallel

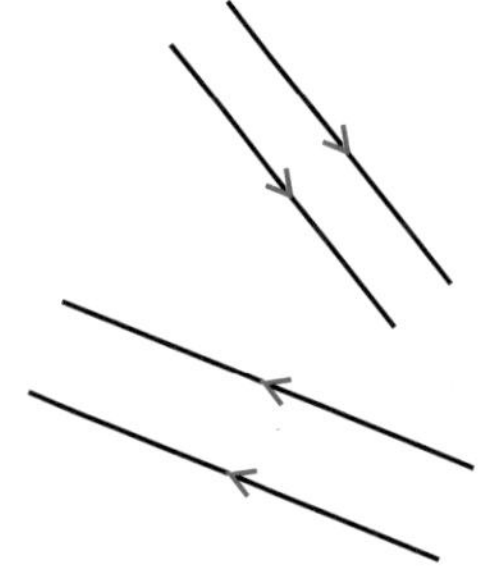

Lines are parallel if they NEVER meet, no matter how far they are extended.

Example: Railway lines are parallel.

Parallel lines are always the same distance apart and they never ***intersect***. Arrows are used to show parallel lines, as shown on the left.

Parallelogram

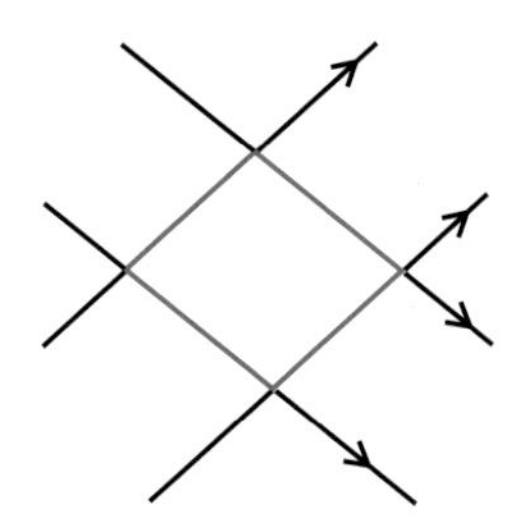

A parallelogram is a ***quadrilateral*** formed by two pairs of ***parallel lines***. In general a parallelogram has:

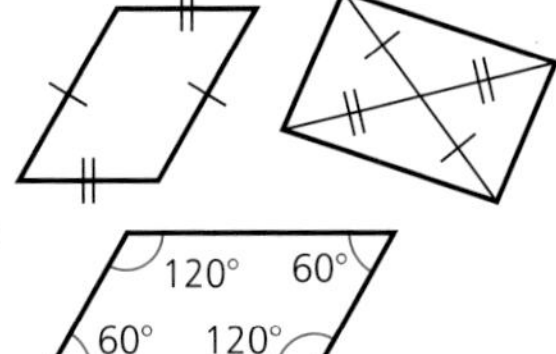

a) NO ***lines of symmetry***;
b) ***rotational symmetry*** of order 2;
c) opposite sides equal;
d) opposite ***angles*** of the figure equal;
e) ***diagonals bisecting*** each other.

Examples (left and right):

Partial variation

Partial variation is where there is a ***constant*** added to a quantity that is ***directly variable*** to another.

The relationship between y and x, where y varies partially as x is $y = ax + b$ where a and b are constants.

Example: The total cost, c, of hiring a car is a fixed charge of \$20 plus a charge of 10 cents for each kilometre driven. If the distance driven is k kilometres, then

$c = 2000 + (10 \times k)$ where c is the cost in cents.

The charge, $10 \times k$, for the distance travelled is ***directly proportional*** to the distance, k, but the total cost, c, is partially propotional to the distance.

The total cost, c and the distance travelled, k are in partial variation.

Pascal's triangle

Pascal's triangle is a pattern of numbers. The triangle pattern starts with

and the general rule for continuing the pattern is – 'add each pair of numbers and write the result underneath; putting 1 at the beginning and 1 at the end'.

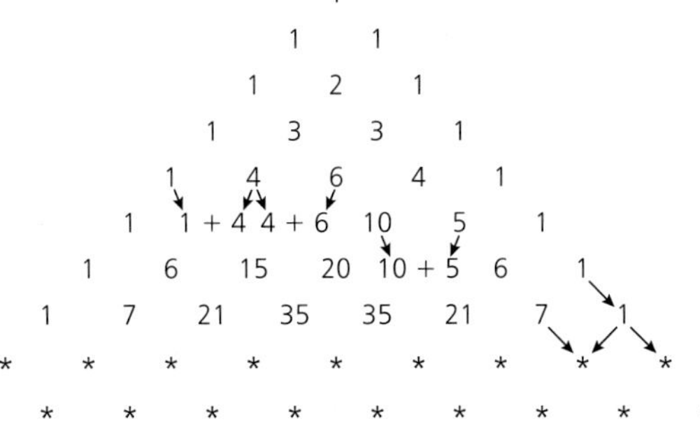

Pentagon

A pentagon is a ***polygon*** bounded by five straight lines and containing five ***angles***.

Examples:

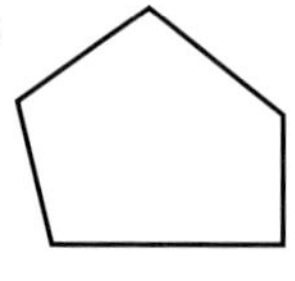

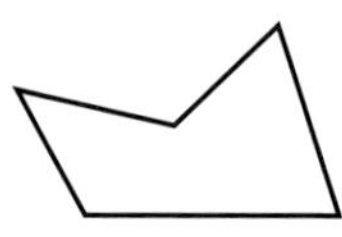

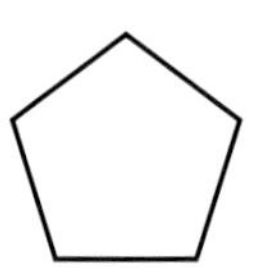

This is a ***regular*** pentagon

Percentage

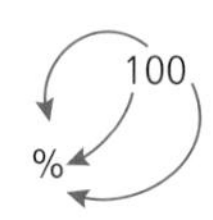

A percentage is a ***fraction*** the ***denominator*** of which is 100.

Examples: $\frac{20}{100} = 20\%$ $\frac{83}{100} = 83\%$ $\frac{22\frac{1}{2}}{100} = 22\frac{1}{2}\%$

NOTE: % is $\frac{1}{100}$ (hundredths) rewritten in a special way.

Percentage error

A percentage error is the ***error*** expressed as a percentage of the actual value.

See **Absolute error**, **Relative error**

Percentile

Percentiles divide the values in a distribution into 100 equal sections. The 20th percentile is the value that 20% of the distribution is less than or equal to. The 50th percentile is the ***median***.

Perfect number

A perfect number is a ***natural number*** which equals the ***sum*** of all its ***factors*** except itself.

Examples:

6 {factors of 6} = {1, 2, 3, 6}
$1 + 2 + 3 = 6$

28 {factors of 28} = {1, 2, 4, 7, 14, 28}
$1 + 2 + 4 + 7 + 14 = 28$

NOTE: After 6 and 28 the next perfect number is 496.

Perfect square

A perfect square is a number or ***expression*** that can be written as the ***product*** of two equal ***factors***.

Example: 9 is a perfect square because $9 = 3 \times 3$

$\frac{4}{25}$ is a perfect square because $\frac{4}{25} = \frac{2}{5} \times \frac{2}{5}$ and

$x^2 + 6x + 9$ is a perfect square because $x^2 + 6x + 9 = (x + 3)^2$.

See **Completing the square**

Perimeter

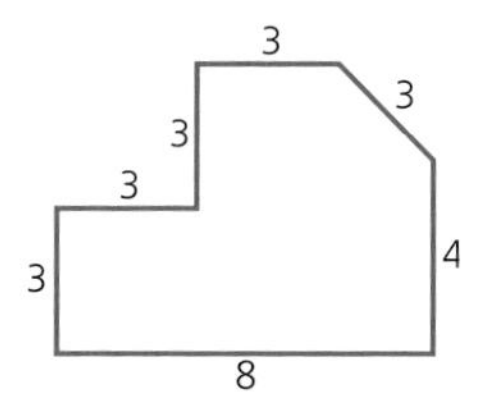

The perimeter of a figure (or shape) is the length of the ***boundary*** or the distance around the outside of the shape.

Example (left):
Perimeter is $3 + 3 + 3 + 3 + 3 + 4 + 8 = 27$

Period

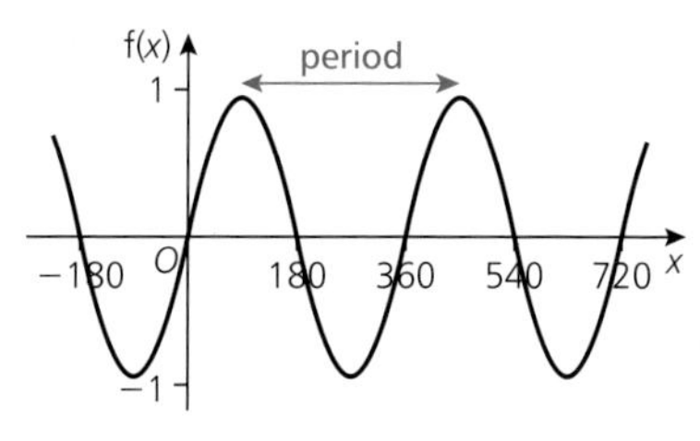

A periodic ***function*** has a repeating pattern. The period is the width of that pattern measured along the *x*-***axis***.

Example: The period of the ***sine*** function is 360°.

Perpendicular

Lines are perpendicular if they cross at ***right-angles***.

Examples:

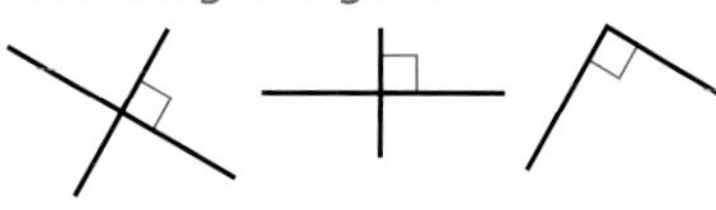

Similarly, ***planes*** are perpendicular if they meet at right-angles. For example, in most rooms a wall and a floor are perpendicular.

Perpendicular bisector

The perpendicular bisector of a ***line*** is ***perpendicular*** to the line and goes through the mid-point of the line.

Phase angle

The phase angle of a ***periodic*** wave is the number of ***units*** of angular measure between a point on the wave and a reference point. The reference point is often the ***origin***.

Example: The phase angle between *P* on the sine wave, $f(x) = \sin x$, and the origin is 120°.

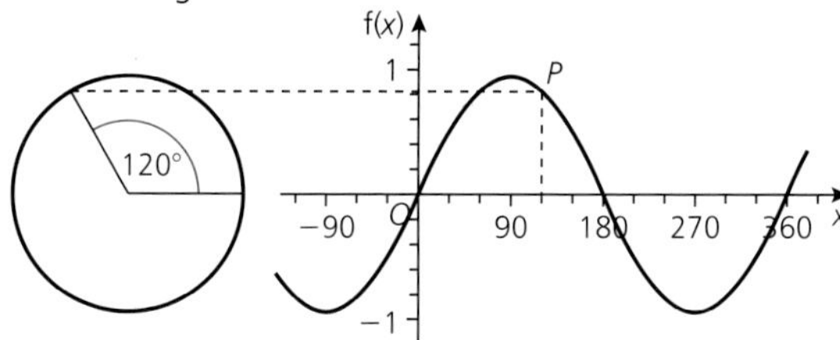

Pi

Pi (or π) is the ***ratio*** of the ***circumference*** of a ***circle*** to its ***diameter***. For ALL circles the ***fraction***

$\frac{\text{circumference}}{\text{diameter}}$ is π or $C = \pi d$

Pi is also used in calculating the ***area*** inside a circle. $A = \pi r^2$

π is approximately	$= 3$	(1 ***significant figure***)
	$= 3.1$	(2 significant figures)
	$= 3.14$	(3 significant figures)

Another good approximation for π is $3\frac{1}{7}$ or $\frac{22}{7}$

$\pi = 3.141\ 592\ 653\ 589\ 79\ \ldots$

Pi is an ***irrational number***.

Examples:

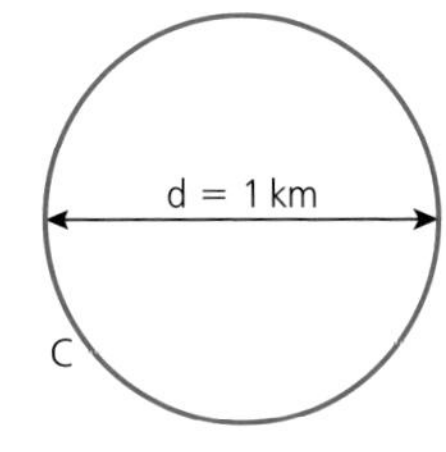

The distance around a circular lake that is 1 km across is

$C = \pi \times d$

$= 3 \times 1$

$= 3$ km (1 S.F.)

40 cm

The area of a circular mirror with ***radius*** 40 cm is

$A = \pi \times r^2$

$= \pi \times 40^2$

$= 3.142 \times 1600$

$= 50.3\ \text{cm}^2$ (3 S.F.)

Pictogram

A pictogram is a way of representing information. It is a kind of ***graph*** used in ***statistics***. It is similar to a ***bar chart***, but made using a motif.

Example:

walk	
bus	
cycle	

represents 1 pupil

Pictogram showing how pupils in form 1B come to school

Pie chart

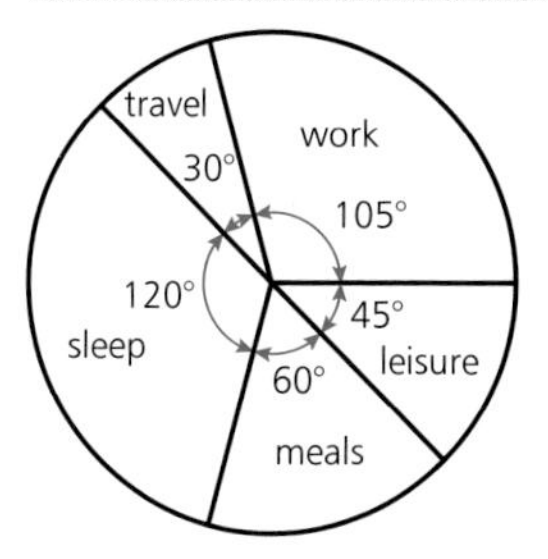

A pie chart is a way of representing information. It is a kind of ***graph*** used in ***statistics***.

Example (left):
Pie chart showing how a person spent their time last Saturday.
15° represent 1 hour.

NOTE: The ***circle*** is divided into ***sectors*** so that the ***areas*** of the sectors represent the ***data***.

Place value

Place value is the position of a ***digit*** in a number that tells you its value.

Example: The digit 4 in the number 2490 is in the hundreds column; the place value is 100, so the digit 4 represents 400.

Plane

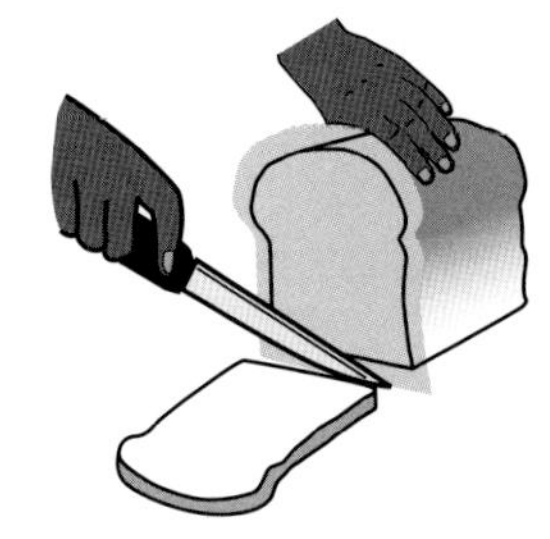

A ***plane*** is a flat surface such as a table-top.

A plane has no thickness and extends for ever into space in all directions.

Example (left): When slicing a loaf of bread, the knife must be kept in one plane to get flat slices.

Plane shape

A plane shape is a shape that can be drawn with all its points in one ***plane***.

Examples: All ***polygons***, such as ***squares*** and ***triangles***, are plane shapes, as are ***circles***.

Plane of symmetry

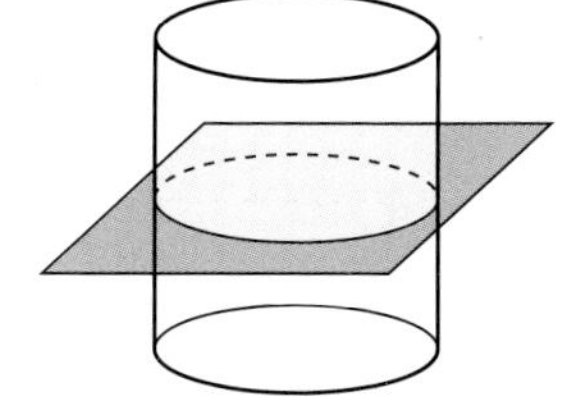

A solid has a plane of symmetry if there is a ***plane*** which can act as a mirror to show the complete shape.

Examples (left and right):
For each solid the plane of symmetry is shown in green.

Polar coordinates

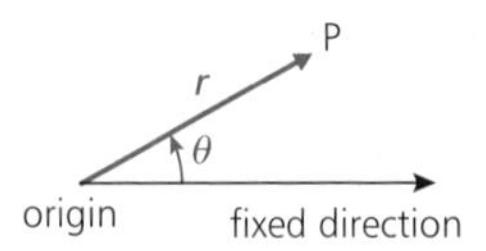

For Cartesian coordinates **see Coordinates**.

Polar coordinates are another way of describing the position of points in a ***plane***. Instead of (x, y), the distance r from the ***origin*** and the ***angle*** θ from a fixed direction are given. The position of a point P is then given by (r, θ) in polar coordinates.

(***Anticlockwise*** angles are indicated by ***positive*** *numbers*.)

Examples:

A is (2, 30°)   B is (1, 90°)
C is (3, 150°)   D is (2, 270°)

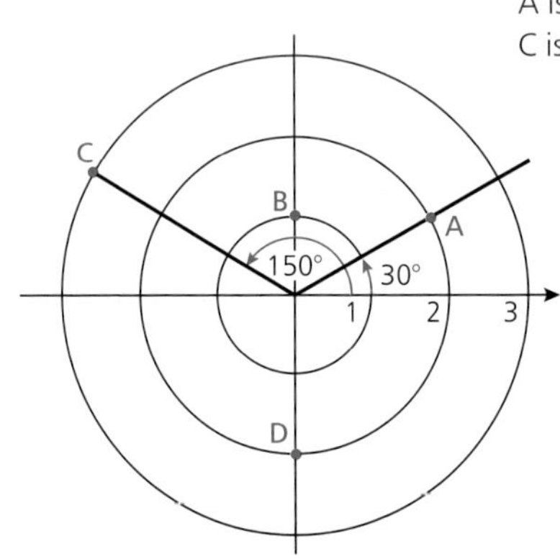

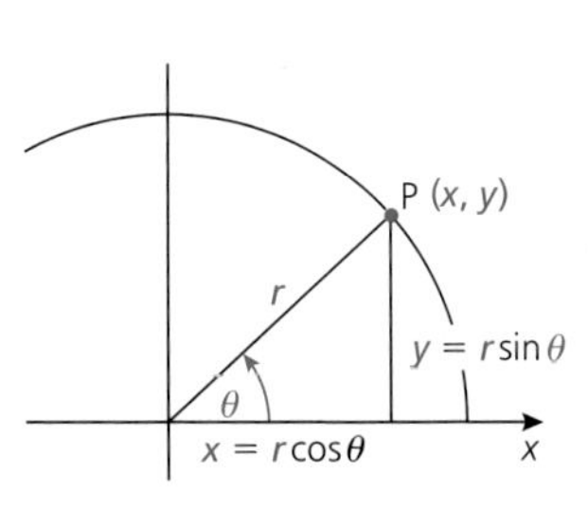

If P is (x, y) in Cartesian coordinates and if P is (r, θ) in polar coordinates then

$x = r\cos\theta$, $y = r\sin\theta$

and $r = \sqrt{x^2 + y^2}$, $\tan\theta = \dfrac{y}{x}$

Polygon

A polygon is a ***plane shape*** bounded by only straight lines.

Examples:

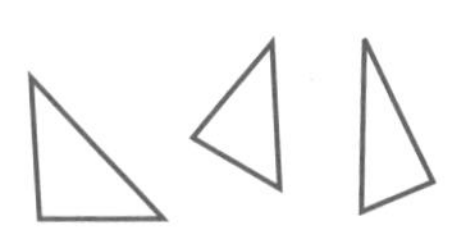

3 sides   triangle

4 sides   quadrilateral

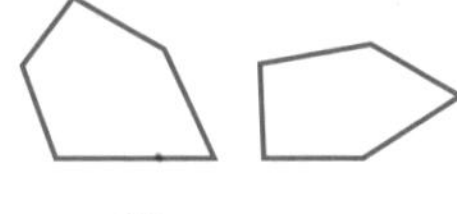

5 sides   pentagon

6 sides   hexagon

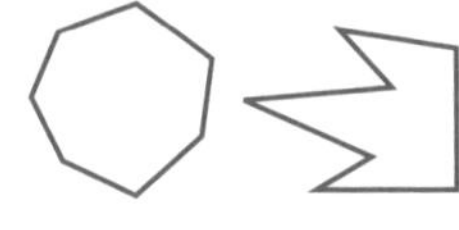

7 sides   heptagon

8 sides   octagon

Polyhedron (plural **Polyhedra**)

A polyhedron is a solid shape with flat sides. The flat sides or ***faces*** are all ***polygons***.

Examples: See **Cube, Cuboid, Dodecahedron, Icosahedron, Octahedron, Prism, Rectangular Prism, Tetrahedron**.

NOTE: There are only these five ***regular*** polyhedra:

tetrahedron

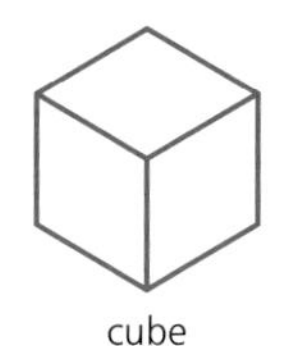

cube

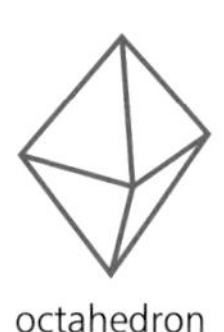

octahedron

dodecahedron

icosahedron

Polynomial

A polynomial is an ***expression*** in one unknown whose ***terms*** are ***constants*** multiplied by ***whole number powers*** of the unknown.

Example: $2x^3 - 4x + 5$ is a polynomial in x.

Position vector

A ***vector*** is used as a position vector when it describes the position of a point in relation to the ***origin***.

See **Displacement**

Positive number

Positive numbers are numbers greater than ***zero*** such as 3, 2.56, 0.00081.

On the ***number line*** any number to the right of zero (as shown by the green line) is a positive number.

−3 −2 −1 0 1 2 3 4 5

Positive ***integers*** work like the ***natural numbers*** and so we do not always use the + sign.

$\{+1, +2, +3, +4, +5, \ldots\} = \{1, 2, 3, 4, 5, \ldots\}$

See **Negative number**

Power of a number

The power of a number is the result of a multiplication using just that number.

Examples:

$2 \times 2 \times 2 = 8$
$2 \times 2 \times 2 \times 2 \times 2 = 32$
$2 \times 2 = 4$

8, 32 and 4 are some powers of 2

2 to the power 3 is $2 \times 2 \times 2$ and is written as 2^3
2 to the power 5 is $2 \times 2 \times 2 \times 2 \times 2$ and is written as 2^5

Similarly

$2^{10} = 2 \times 2 \times 2 \times 2 \times 2 \times 2 \times 2 \times 2 \times 2 \times 2 = 1024$

$3^4 = 3 \times 3 \times 3 \times 3 = 81$

$5^3 = 5 \times 5 \times 5 = 125$

7
7
area
49

Note: A number 'to the power 2' is said to be '***squared***'.

7^2 is 7 squared
$= 7 \times 7$
$= 49$

It is equal to the ***area*** of a ***square*** of side 7.

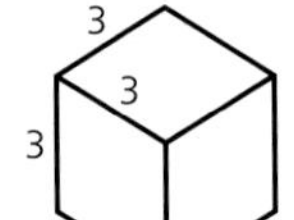

A number 'to the power 3' is said to be '***cubed***'.

3^3 is 3 cubed
$= 3 \times 3 \times 3$
$= 27$

It is equal to the ***volume*** of a ***cube*** of side 3.

Prime factor

Prime factors of a number are ***factors*** of the number which are also ***prime numbers***.

Example:

{factors of 24} = {1, 2, 3, 4, 6, 8, 12, 24}
(2, 3: prime numbers)

{prime factors of 24} = {2, 3}

Note: We can write any number as a ***product*** of prime factors.

Examples:

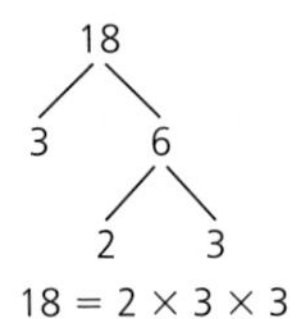

$18 = 2 \times 3 \times 3$

60
6
10
2
3
2
5

$60 = 2 \times 2 \times 3 \times 5$

Prime number

A prime number has only two ***factors***, itself and 1.

Examples:

$17 = 17 \times 1$ $\quad$ $13 = 13 \times 1$

17 and 13 cannot be written as ***products*** of any other ***natural numbers***.

{prime numbers} = {2, 3, 5, 7, 11, 13, 17, 19, 23, 29, 31, 37, 41, 43, ...}

A prime number is a number that is not a ***rectangle number***.

NOTE: The number 1 is not a prime number and it is not a rectangle number.

Principal

The principal is the sum of money that is initially invested or borrowed on which interest is calculated.

Prism

A prism is a ***polyhedron*** with the same shape along its length.

Examples:

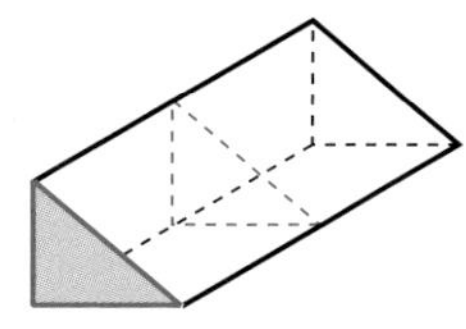

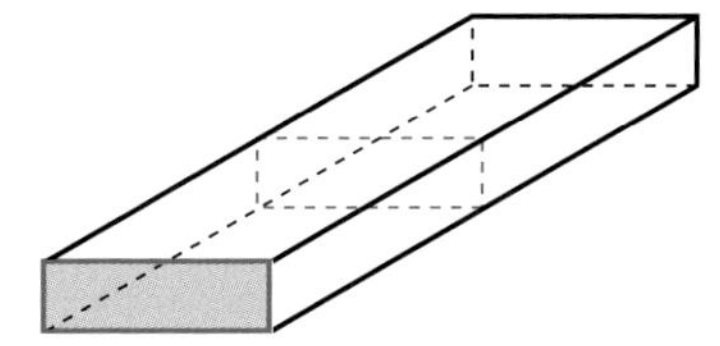

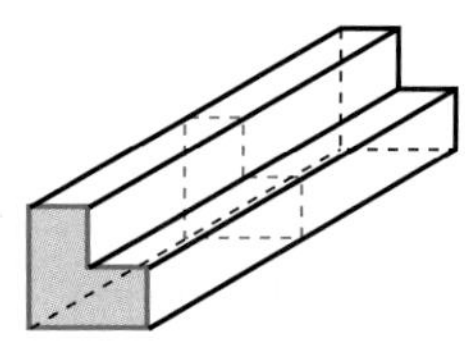

Probability

Probability is a measure of how likely an event is. The probability of an event is a number between 0 and 1.

Examples:

Probability of a man having two heads is 0.

Probability of rolling a three with one die is $\frac{1}{6}$ or 0.16666...

Probability of a newborn baby being a boy is $\frac{1}{2}$ or 0.5.

Probability of not picking a spade from a new pack of cards is $\frac{3}{4}$.

Probability of you dying in the next 200 years is 1.

Mathematically, probability of an event S

$$= \frac{\text{number of ways S can happen}}{\text{number of possible outcomes}}$$

provided that the possible outcomes are all ***equally likely***.

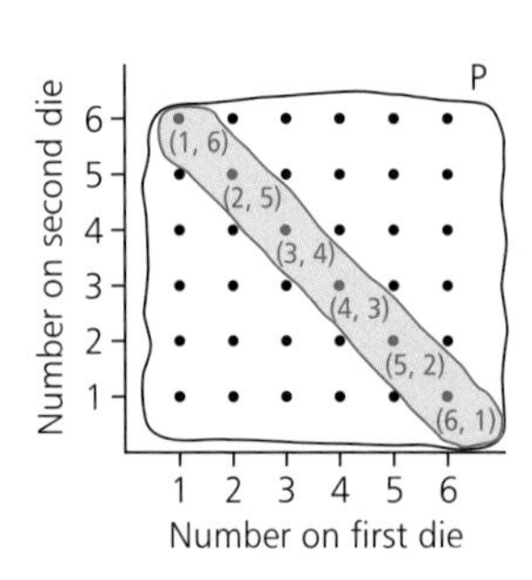

Example (left):
S = {a total score of seven when two dice are rolled}
P = {all possible number pairs when two dice are rolled}
S = {(1, 6), (2, 5), (3, 4), (4, 3), (5, 2), (6, 1)}

number in **S** is 6

P = {all 36 points on the graph}
number in **P** is 36

so, the probability of scoring seven when two dice are rolled is $\frac{6}{36} = \frac{1}{6}$

Probability space

A probability space is all the possible outcomes. In the example for ***probability*** above, the 36 points shown by the set P is the probability space for this situation.
See **Equally likely, Independent event, Mutually exclusive**

Product

The product of two or more numbers is the result of multiplying them together.

Examples:
The product of 6 and 7 is $6 \times 7 = 42$
The product of 2 and 5 is $2 \times 5 = 10$

The same word is used for the result of multiplying two ***matrices*** together.

Profit

Profit is the ***difference*** between the price that an object is sold for and the price that it was bought for.

Proportion

Two quantities are in proportion when corresponding pairs are always in the same ***ratio***.

Example: The number of pens, n, bought in a shop at a cost of C pence is shown in this table.

Number of pens, n	1	2	3	4	5	10	20
Cost of pens, C	2	4	6	8	10	20	40

The ratios 2 to 4, 5 to 10, 20 to 40, etc., are equal, so the quantities n and C are in proportion.

See **Direct proportion**

Pyramid

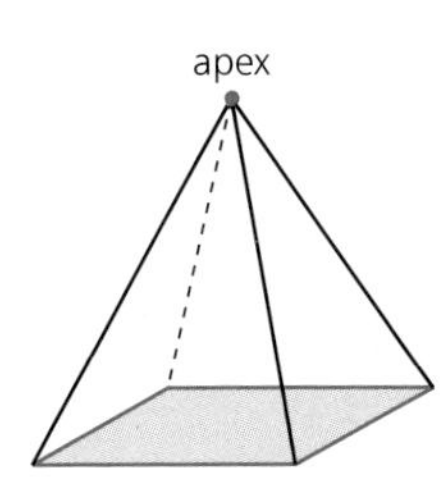

A pyramid is a ***polyhedron*** with a ***polygon*** for a base and all other faces meeting at one vertex called the apex.

Examples: The bases are shaded green and some apexes are shown by a green dot.

The pyramid on the left is a ***square***-based pyramid

This is a ***triangular***-based pyramid, it is the same solid as a ***tetrahedron***.

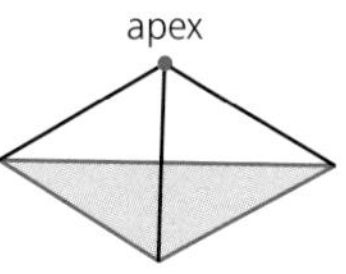

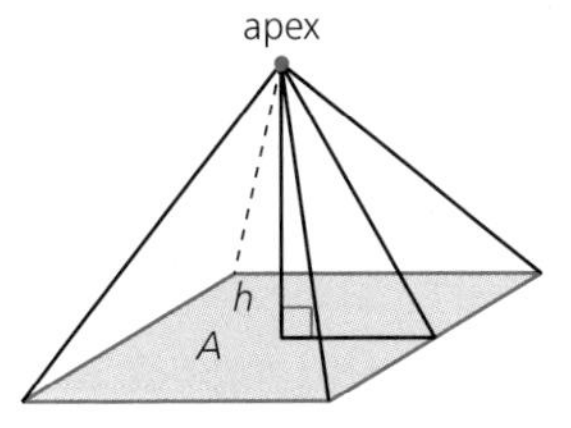

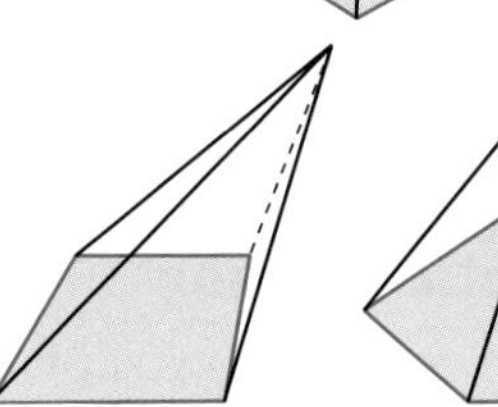

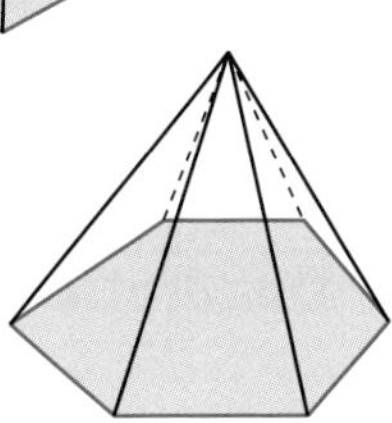

The ***volume*** of ANY pyramid is given by

Volume $= \frac{1}{3} \times$ base area $\times$ ***perpendicular*** height of the apex above the base

$V = \frac{1}{3}Ah$

Pythagoras

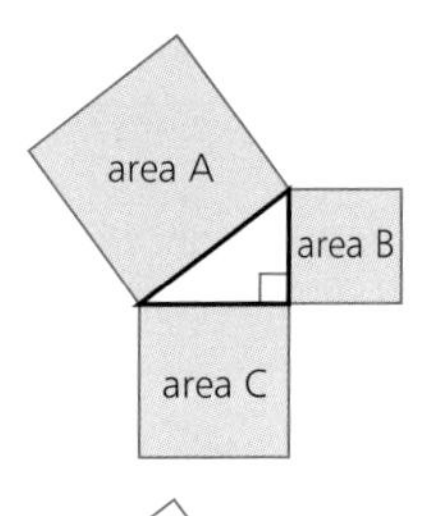

Pythagoras' theorem says that in a ***right-angled triangle***, the ***area*** of a ***square*** on the longest side is equal to the ***sum*** of the areas of squares drawn on the two smaller sides.

Area A = Area B + Area C

OR

$p^2 = q^2 + r^2$

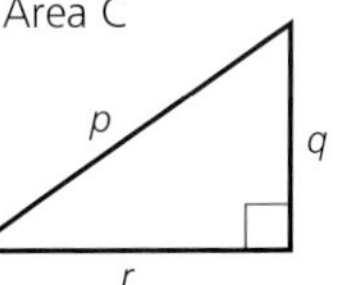

Examples:

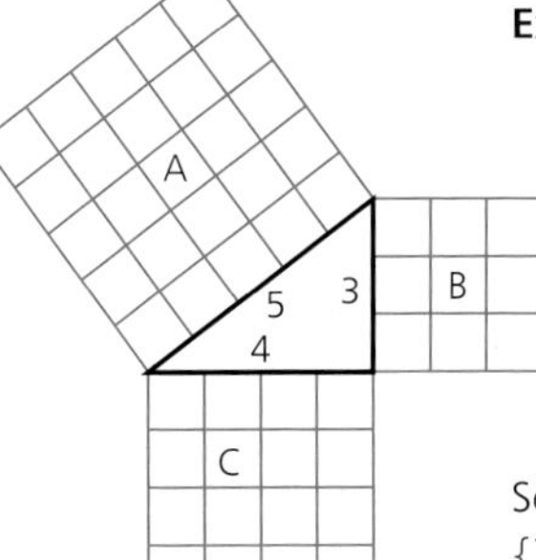

A = B + C

25 = 9 + 16

OR

$5^2 = 3^2 + 4^2$

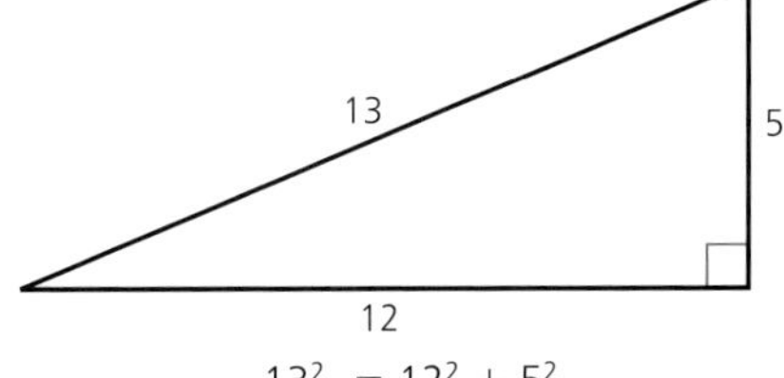

$13^2 = 12^2 + 5^2$

$169 = 144 + 25$

Some ***sets*** of three numbers that fit this pattern are

{3, 4, 5} {5, 12, 13} {8, 15, 17}

{7, 24, 15} {20, 21, 29} {12, 35, 37}

Quadratic equation

A quadratic equation in x is an ***equation*** involving x to the ***power*** 2 but no higher powers of x.

The general form of a quadratic equation is
$ax^2 + bx + c = 0$

Examples:
$3x^2 = 12$ $x^2 + 1 = 26$ $x^2 - 2x + 1 = 0$

When a quadratic equation has ***solutions*** and the equation is written in the general form, the solutions can be found using the ***formula***

$$x = \frac{-b \pm \sqrt{b^2 - 4ac}}{2a}$$

Quadratic function

A quadratic ***function*** is a ***relation*** of the form
$x \rightarrow ax^2 + bx + c$ where $a \neq 0$

Examples:
$x \rightarrow 2x^2$
$x \rightarrow x^2 + 3$
$x \rightarrow 5x^2 + x - 13$
$x \rightarrow 9 - x^2$

A quadratic function with a ***positive*** x^2 term has a ***minimum*** *value*.
A quadratic function with a ***negative*** x^2 term has a ***maximum*** *value*.

Quadrilateral

A quadrilateral is a ***plane shape*** bounded by four straight ***lines***.

Examples:

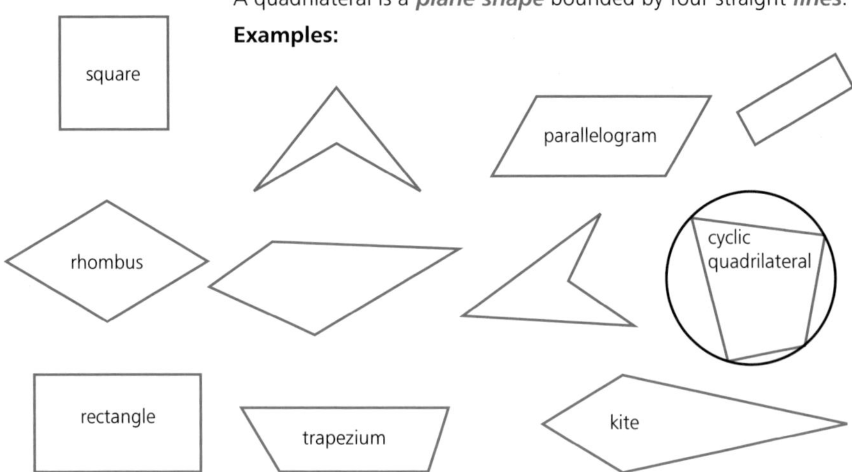

Quartile

When there is a large amount of ***data*** some idea of their spread is given by the quartiles. These are the values for which $\frac{1}{4}$ or $\frac{1}{2}$ or $\frac{3}{4}$ of the data is less. We usually use a ***cumulative frequency*** diagram to find the quartiles.

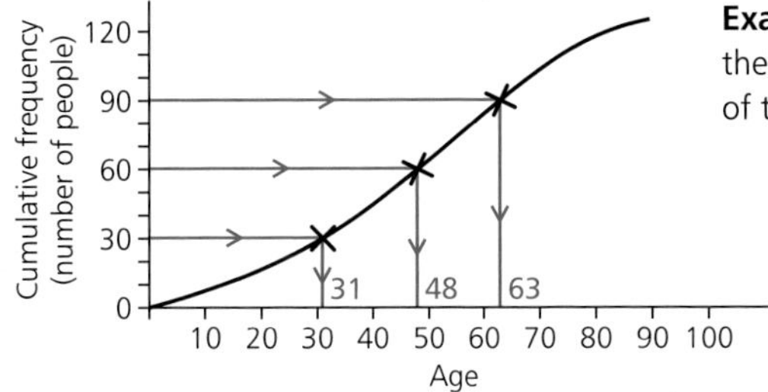

Example (left): In a village 120 people are asked their age. The first quartile is 31 years because $\frac{1}{4}$ of the people are 31 years or less.

The second quartile is 48 years because $\frac{1}{2}$ of the people are 48 years or less.
The third quartile is 63 years because $\frac{3}{4}$ of the people are 63 years or less.
The second quartile is the same as the ***median***.

Quartile deviation

The quartile deviation is half the ***difference*** between the first and third ***quartiles***. It is also called the ***semi-interquartile range***.

Quotient

The quotient is the result you get when you do a ***division***.

Example:
In the division $6\overline{)4218}$ with result 703 — 703 is the quotient.

R

Radian

A radian is a measure of ***angle***. 1 radian $= \frac{180}{\pi}$ degrees.

Radius

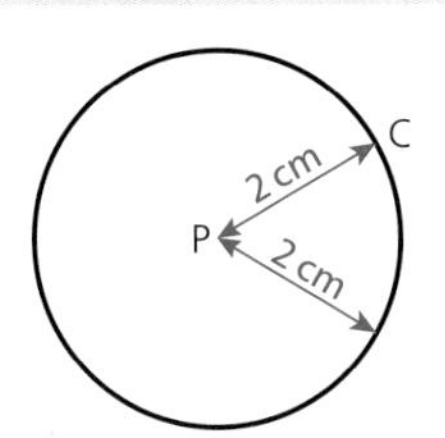

The radius of a ***circle*** is the distance from the centre of the circle to a point on the circle.

Example (left):
All the points on the circle C are 2 cm from the centre P.

Range

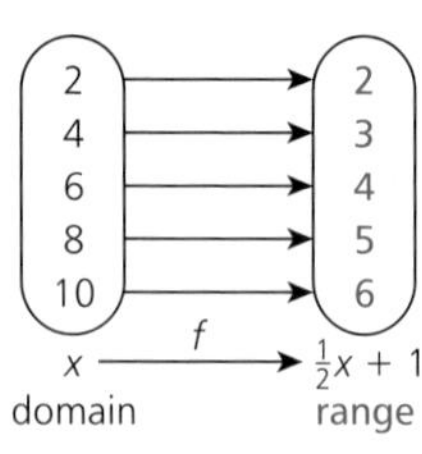

First meaning

(***Mappings***) The range of a ***function*** is the ***set*** of numbers onto which the ***domain*** is ***mapped***. The range is the same as the ***image*** set.

Example (left):

{2, 4, 6, 8, 10} is the domain {2, 3, 4, 5, 6} is the range

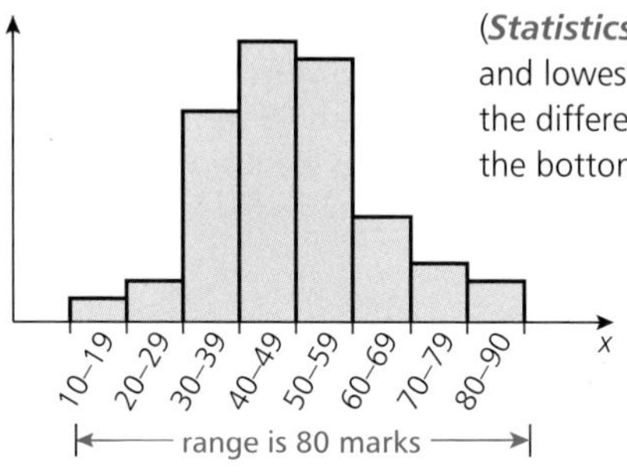

Second meaning

(***Statistics***) The range is the ***diffference*** between the highest and lowest value in the ***data***. For grouped data the range is the difference between the top end of the highest group and the bottom end of the lowest group.

Example:

135 children took an examination.
The marks were grouped. The top end of the highest group is 90 and the bottom end of the lowest group is 10.
The range is $90 - 10 = 80$ marks

Rate

Rate is one quantity measured against one ***unit*** of another quantity.

Example: An interest rate of 5% per annum means the interest is 5% each year.

Rate of change

Rate of change measures the change in one quantity for one ***unit*** increase of another quantity.

Example: ***Velocity*** is the change in ***displacement*** in one unit of time so velocity is the rate of change of displacement with respect to time.

Ratio

LEAGUE TABLE		
	Won	Lost
Chelsea	5	11
Newcastle	20	6
Liverpool	17	10
West Ham	14	14
Arsenal	12	19

A ratio is used to compare two or more quantities.

Examples:

A youth club has 30 boy members and 40 girl members.
The ratio of boys to girls is 30 to 40

A football team has won 17 matches and lost 10 matches.
The ratio of wins to losses is 17 to 10

With each ratio we think of a ***fraction*** that goes with it.

Examples:

The ratio 5 to 7 and the fraction $\frac{5}{7}$

The ratio 15 to 20 and the fraction $\frac{15}{20}$ or $\frac{3}{4}$

NOTE: Ratios are equal when their fractions are ***equivalent***.

Rational number

A rational number is one that can be written as a ***fraction*** whose numerators and denominators are ***integers***.

The ***set*** of rationals is $\left\{\frac{p}{q}\right.$ such that p and q are integers, $q \neq 0$, and p and q have no common ***factor***$\left.\right\}$.

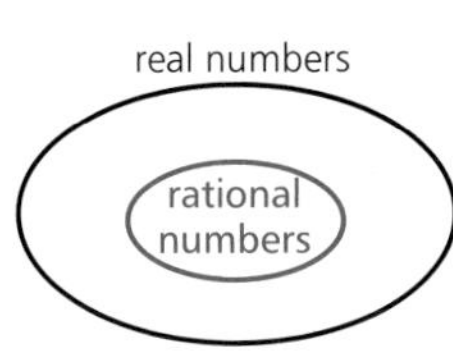

Examples: $\frac{2}{3}, \frac{-7}{8}, \frac{15}{19}, \frac{8}{1}, \frac{1}{13}, \frac{-42}{5}$ are examples of rational numbers.

See **Irrational number**

The set of rational numbers is a ***subset*** of the set of ***real numbers***.

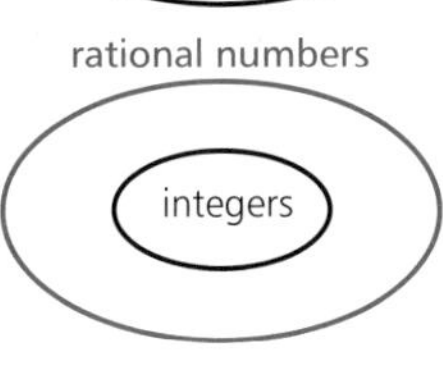

The set of integers is a subset of the set of rational numbers.

NOTE: A rational number can always be written as a decimal that either ***terminates*** or ***recurs***.

Rationalising the denominator

Rationalising the denominator means converting a ***fraction*** whose ***denominator*** contains an ***irrational number*** to a fraction whose denominator is a ***rational number***.

Example: $\frac{2}{\sqrt{3}} = \frac{2}{\sqrt{3}} \times \frac{\sqrt{3}}{\sqrt{3}} = \frac{2\sqrt{3}}{3}$

Ray

A ray is a ***line*** with one end and extending indefinitely in the other direction.

Real number

The ***set*** of real numbers is the set of ALL numbers that can be written as a ***decimal***.

The set of real numbers is the ***union*** of the set of ***rational numbers*** and the set of ***irrational numbers***.

See **Continuous**

NOTE: An example of a number that is NOT real is $\sqrt{-1}$. Numbers such as $\sqrt{-1}$, $\sqrt{-7}$, and $\sqrt{-16}$ are called imaginary numbers.

Reciprocal

The reciprocal of a number x is the number $\frac{1}{x}$

Examples:

The reciprocal of 2 is $\frac{1}{2}$

The reciprocal of 8 is $\frac{1}{8}$

The reciprocal of $\frac{1}{3}$ is 3

The reciprocal of $\frac{4}{7}$ is $\frac{7}{4}$

The reciprocal of $\frac{10}{3}$ is $\frac{3}{10}$

The reciprocal of a number is the same as the ***multiplicative inverse*** of the number.

Rectangle

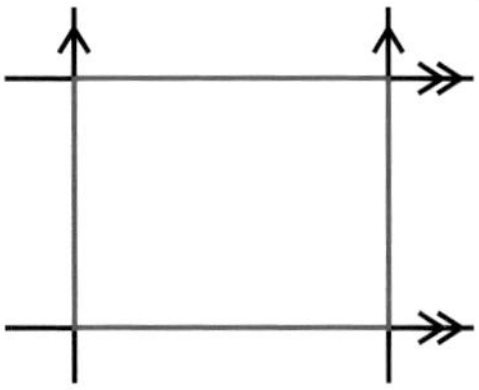

A rectangle is a ***quadrilateral*** formed by two pairs of ***parallel lines*** crossing at ***right-angles***.

Examples:

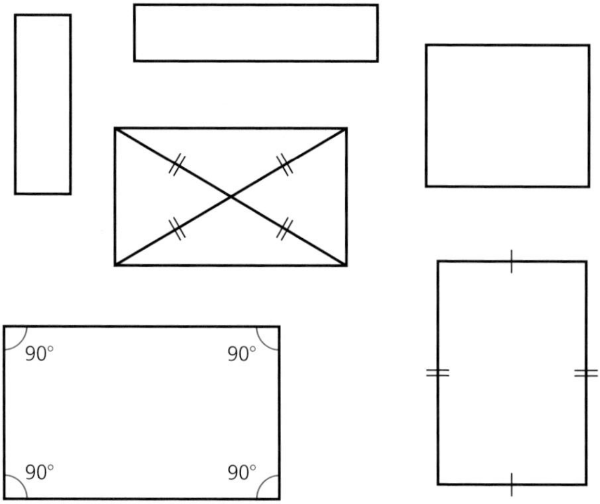

90° 90°
90° 90°

See **Square**

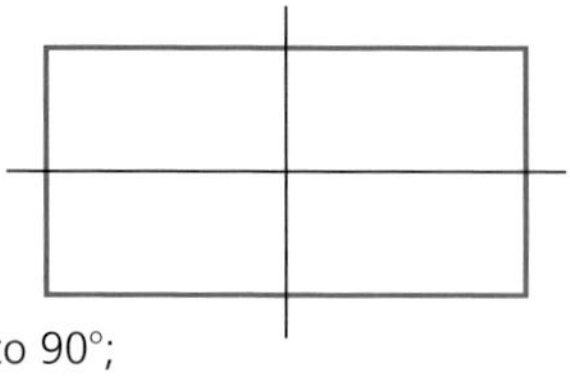

In general a rectangle has:

a) two ***lines of symmetry***;
b) ***rotational symmetry*** of ***order*** 2;
c) opposite sides equal;
d) all four ***angles*** of the figure equal to 90°;
e) ***diagonals*** equal and ***bisecting*** each other.

Rectangle number

A rectangle number is a number that can be shown as a pattern of dots in the shape of a ***rectangle***.

Examples:

$6 = 2 \times 3$

• • •
• • •

$16 = 4 \times 4$ or

• • • •
• • • •
• • • •
• • • •

$16 = 2 \times 8$

• • • • • • • •
• • • • • • • •

$21 = 3 \times 7$

• • • • • • •
• • • • • • •
• • • • • • •

{rectangle numbers} = {4, 6, 8, 9, 10, 12, 14, 15, 16, 18, ...}

See **Prime number**, **Square number**, **Composite number**

NOTE: The number 1 is not a rectangle number and it is not a ***prime number***.

Rectangular prism

A rectangular prism is a ***prism*** with the shape of a ***rectangle*** along its length.

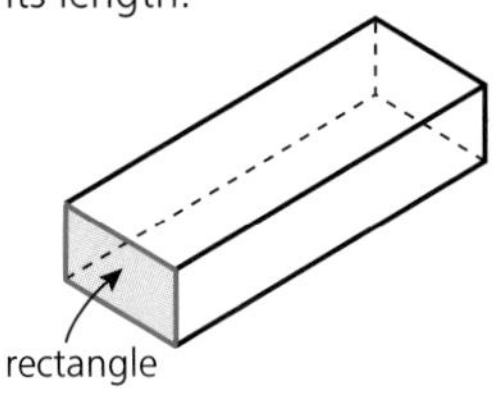

See **Cuboid**

Recurring decimal

A recurring decimal has one ***digit***, or a group of digits, that is repeated endlessly.

Examples:

$\frac{1}{3} = 0.333333\ldots$ $\frac{6}{11} = 0.545454\ldots$

$\frac{1}{7} = 0.142857142857142857\ldots$

We shorten these answers like this:

$\frac{1}{3} = 0.\dot{3}$ $\frac{6}{11} = 0.\dot{5}\dot{4}$ $\frac{1}{7} = 0.\dot{1}4285\dot{7}$

NOTE: 0.4040040004000040000040000004...
has a kind of repeating pattern but it is NOT a recurring ***decimal*** (it is an ***irrational number***).

Re-entrant

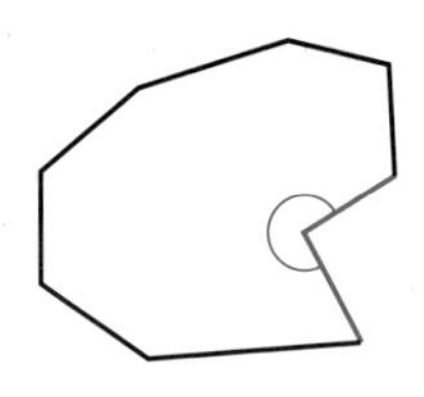

A ***polygon*** which is not ***convex*** is said to be re-entrant.

Examples:

The re-entrant part is shown in green.

In a re-entrant polygon, one or more of the ***angles*** of the polygon is ***reflex***.

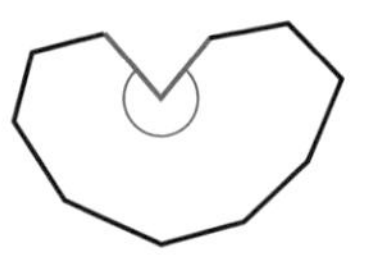

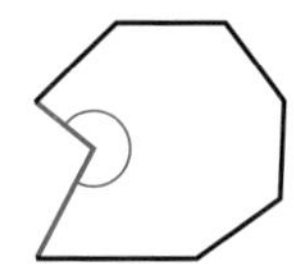

Reflection

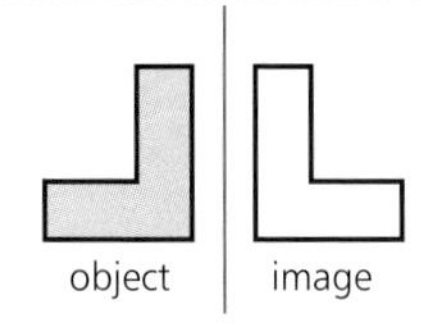

Reflection is a way of ***transforming*** a shape as a mirror does. In a ***plane***, the result of giving an ***object*** a reflection in a ***mirror line*** is called its mirror ***image***.

Example (right):

If P and P′ are ***corresponding points*** then the mirror line m is a ***mediator*** of PP′.

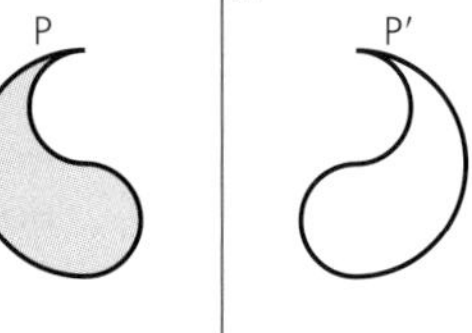

Reflection matrices

Some of the simpler reflections that can be represented as a 2 by 2 ***matrix*** are

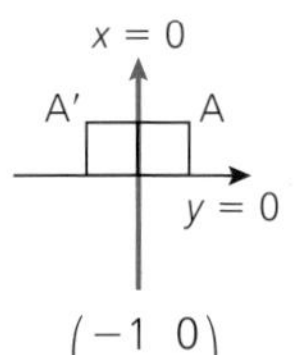

$\begin{pmatrix} -1 & 0 \\ 0 & 1 \end{pmatrix}$

$x = 0$ as mirror line

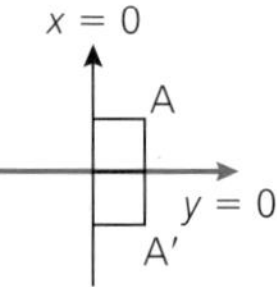

$\begin{pmatrix} 1 & 0 \\ 0 & -1 \end{pmatrix}$

$y = 0$ as mirror line

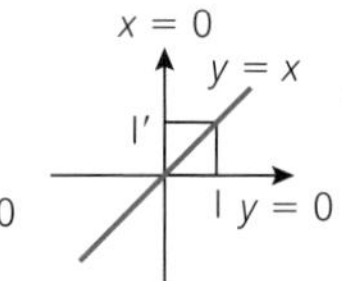

$\begin{pmatrix} 0 & 1 \\ 1 & 0 \end{pmatrix}$

$y = x$ as mirror line

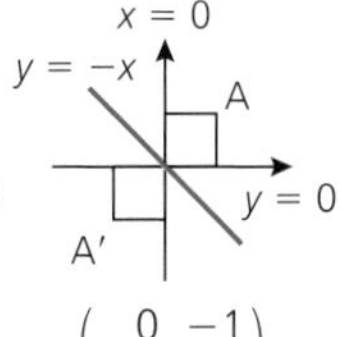

$\begin{pmatrix} 0 & -1 \\ -1 & 0 \end{pmatrix}$

$y = -x$ as mirror line

Reflex angle

A reflex angle is an ***angle*** that is more than 180° and less than 360°.

Examples:

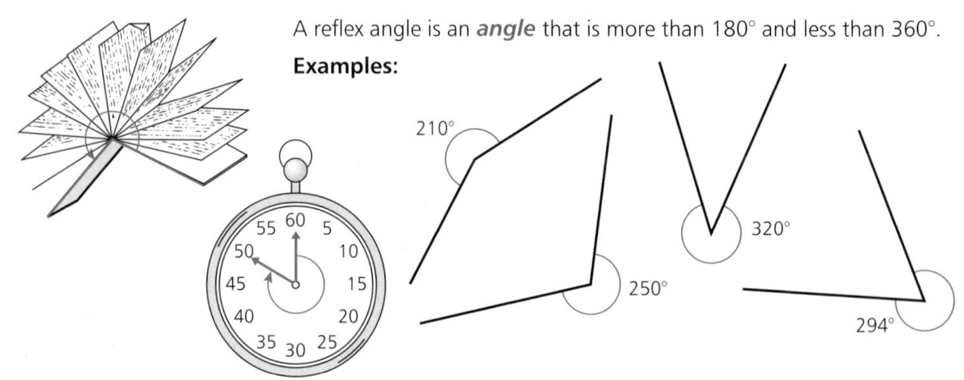

Region

A region is a part of a surface.

Regions on a graph

On a ***graph*** a straight ***line*** divides the ***Cartesian plane*** into two regions.

Examples:

The straight line $x = 4$ (*left*) divides the plane into two regions. The points in the region shaded green are described by the ***inequality*** $x > 4$

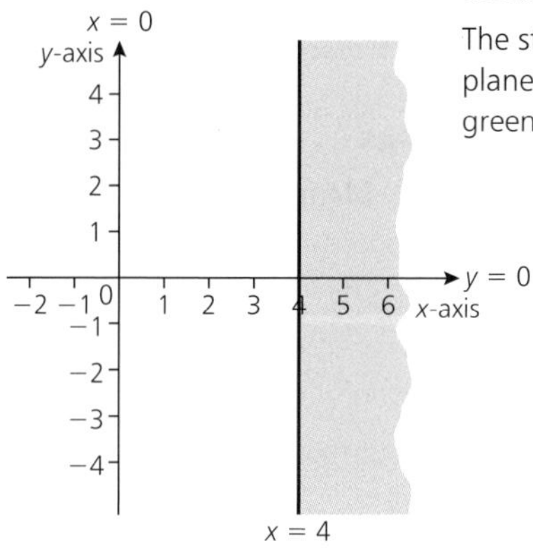

The straight lines $y = -1$ and $y = 2$ (*below*) divide the plane into three regions. The points in the region shaded green are described by the inequality $-1 < y < 2$

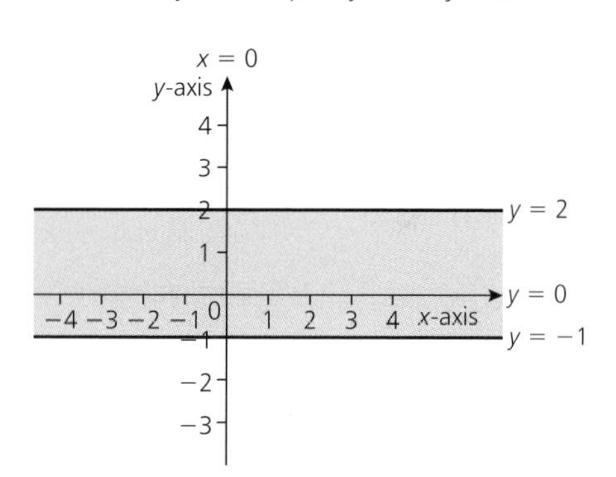

Regular

Regular polygon

A regular ***polygon*** has all its sides the same length and all its ***angles*** the same size.

Examples:

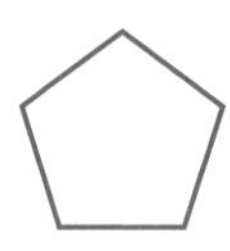

Regular polyhedron

A regular ***polyhedron*** is a polyhedron with ***identical*** regular polygons for all of its ***faces***.

Relation

Relation in general

A relation is a way of connecting ***sets*** of things such as numbers or people.

Examples:

'—is the mother of—'

'—is a factor of—'

'—is 5 more than—'

'—is greater than—'

Relations can be shown by an ***arrow diagram***.

Relations which are mappings

A ***mapping*** is a special relation. See **Mapping**.

Most mathematical relations are mappings and are called either mappings or ***functions***.

See **Functions**.

The same mathematical relation can be shown in various ways.

Example: Here is just ONE relation but shown in five different ways:

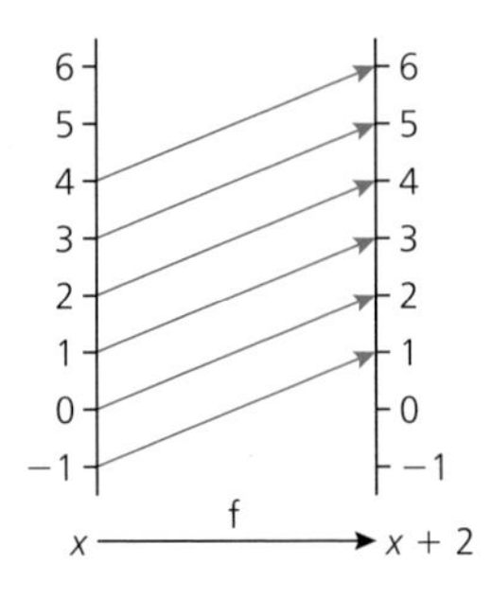

This is a ***mapping diagram***

This is a ***graph***

This is a set of ***ordered pairs***

$\{(-1, 1), (0, 2), (1, 3), (2, 4), (3, 5), (4, 6)\}$

$x \rightarrow x + 2$ This is a relation or mapping.

$y = x + 2$ This is an ***equation***.

Relative error

The relative error is the ***fraction*** $\frac{\textbf{absolute error}}{\text{actual value}}$. It does not have ***units***.

See **Error, Percentage error**

Rhombus

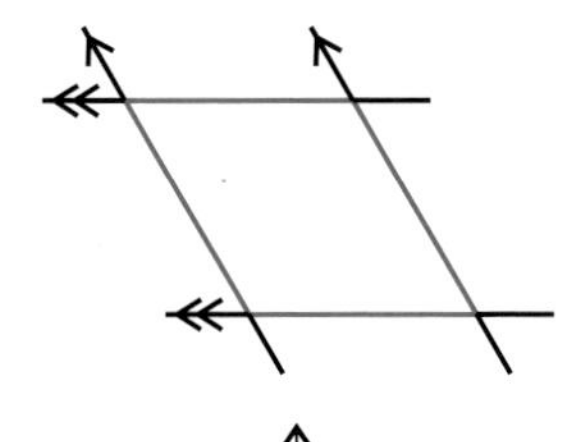

A rhombus is a ***quadrilateral*** formed with four equal sides.

Examples:

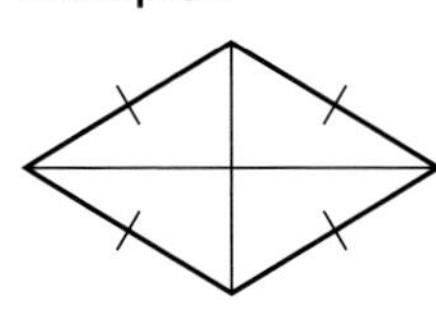

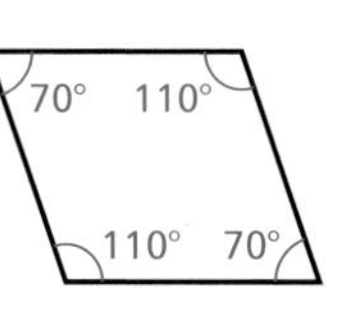

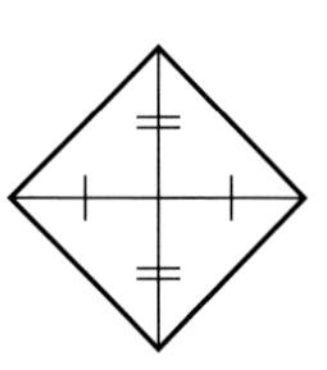

55°
55°
35° 35°

In general a rhombus has:

a) two ***lines of symmetry*** (its ***diagonals***);
b) ***rotational symmetry*** of ***order*** 2;
c) opposite sides ***parallel***;
d) opposite ***angles*** of the figure equal;
e) diagonals ***bisecting*** each other at ***right-angles***;
f) diagonals bisecting the angles of the rhombus.

Right-angle

A right-angle is a quarter of a turn.
It is measured as an ***angle*** of 90°.

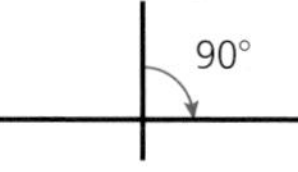

This very important angle is usually shown by a little box as illustrated in the diagram.

Examples:

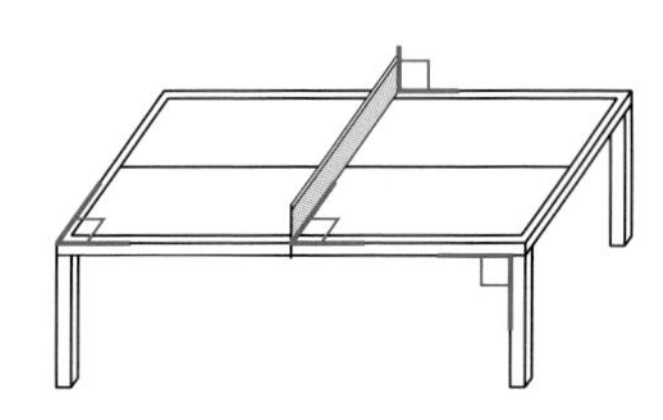

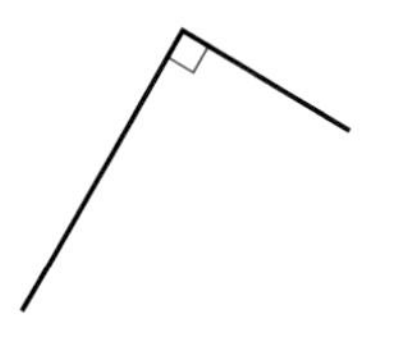

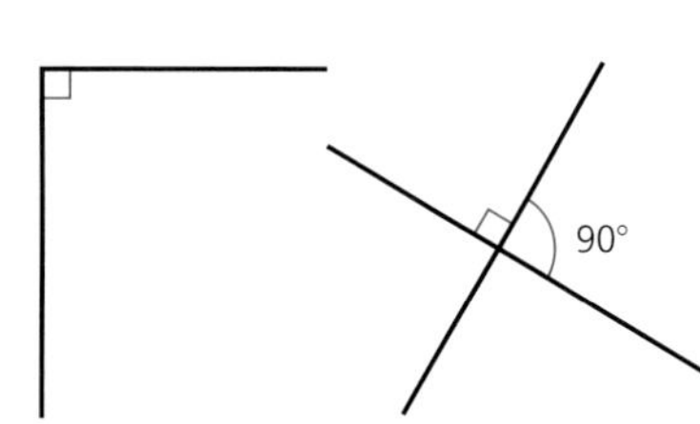

Right-angled triangle

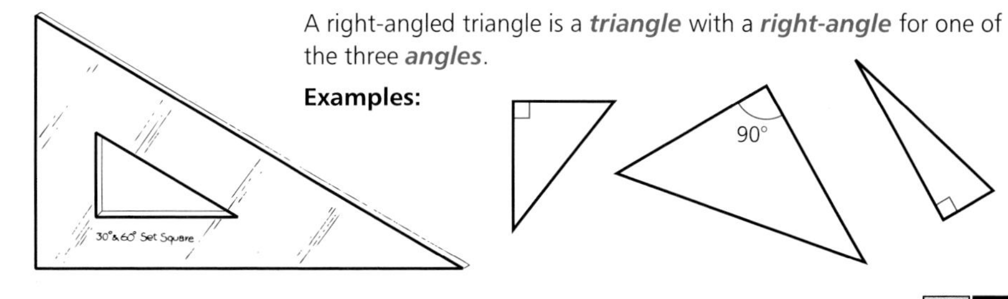

A right-angled triangle is a ***triangle*** with a ***right-angle*** for one of the three ***angles***.

Examples:

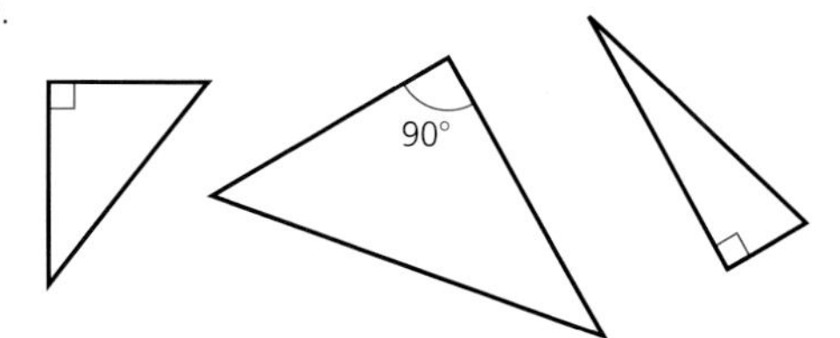

A right-angled ***isosceles*** triangle also has two equal sides.

Examples:

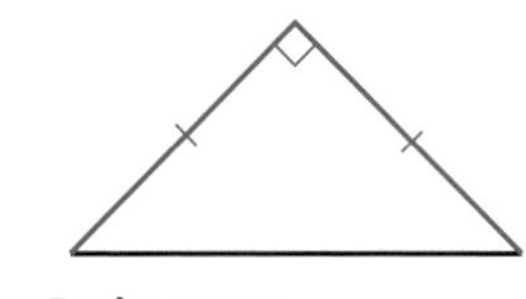

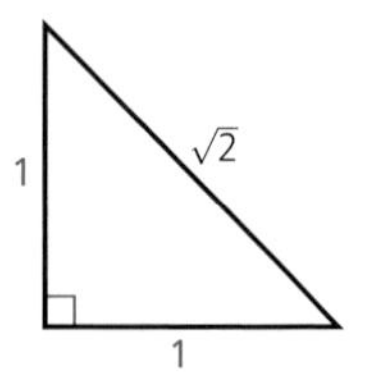

See **Pythagoras**

Right pyramid

A right ***pyramid*** has its ***vertex*** over the centre of its base.

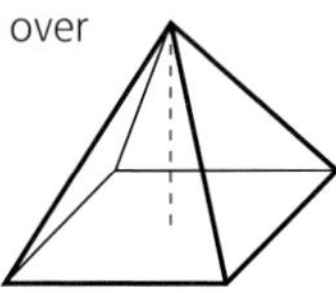

Root

The root(s) of an ***equation*** are the ***solution(s) of the equation***. The roots of a ***function***, $f(x)$, are the values of x for which $f(x) = 0$.

Example: The roots of the equation $(x - 2)(x - 3) = 0$ are 2 and 3.

Rotation

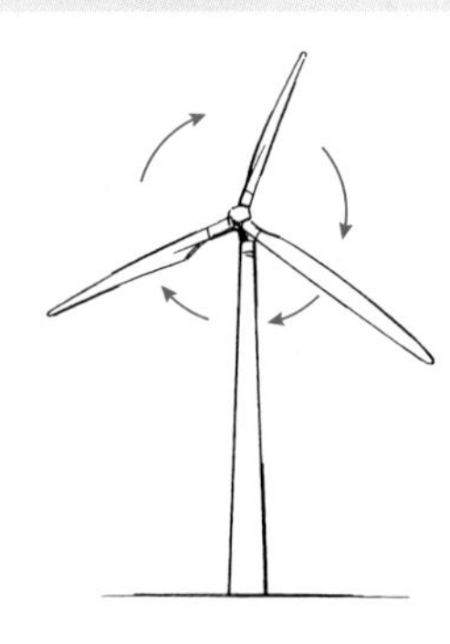

A rotation is a ***transformation*** in which every point turns through the same ***angle*** about the same centre, called the ***centre of rotation***. It is the one point that is ***invariant***.

Example: In this rotation each line a, b and c has turned through an angle of 60° about the centre O.
A rotation of 60° about O gives the green P as the ***image*** of the black P.

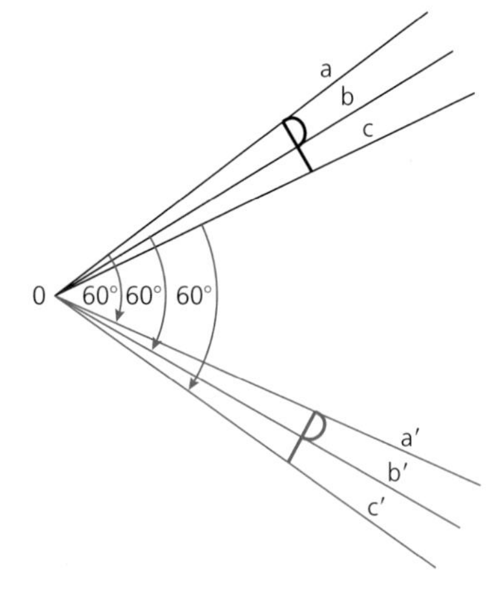

Rotation matrices

Some of the simpler rotations that can be represented as a 2 by 2 ***matrix*** are those rotations about the ***origin***.

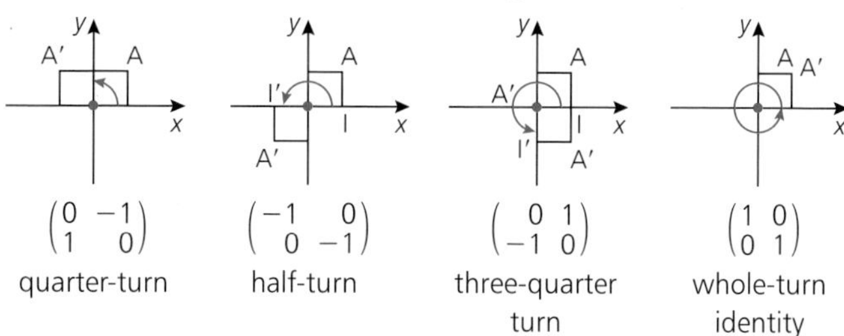

$\begin{pmatrix} 0 & -1 \\ 1 & 0 \end{pmatrix}$ quarter-turn

$\begin{pmatrix} -1 & 0 \\ 0 & -1 \end{pmatrix}$ half-turn

$\begin{pmatrix} 0 & 1 \\ -1 & 0 \end{pmatrix}$ three-quarter turn

$\begin{pmatrix} 1 & 0 \\ 0 & 1 \end{pmatrix}$ whole-turn identity

The matrix equivalent to a rotation through an angle θ

is $\begin{pmatrix} \cos\theta & -\sin\theta \\ \sin\theta & \cos\theta \end{pmatrix}$

Rotational symmetry

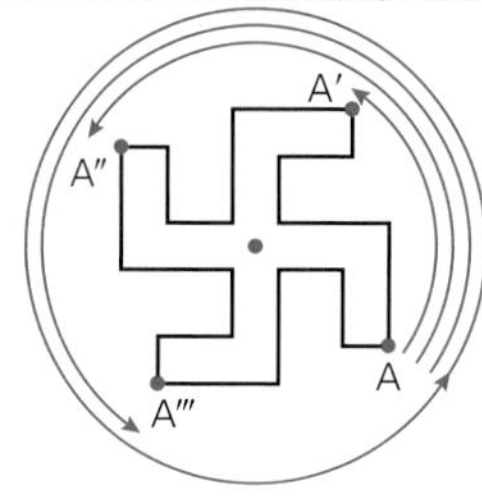

A shape has rotational ***symmetry*** when it can be ***rotated*** about a point and still look the same.

Examples: The shape on the left has rotational symmetry. There are four rotations about O which will leave the shape looking the same.

A rotation of 90° about O will map, for example, A onto A′.

A rotation of 180° about O will map, for example, A onto A″.

A rotation of 270° about O will map, for example, A onto A‴.

A rotation of 360° about O will map, for example, A onto A.

Order of rotational symmetry

The order of rotational symmetry for a shape is the number of possible rotations in one revolution that leave the shape looking the same.

Examples:

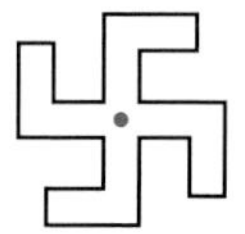

For this shape the order of rotational symmetry is 4.

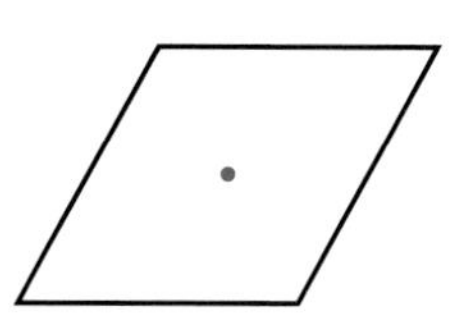

For this shape the order of rotational symmetry is 2

For this fan the order of rotational symmetry is 3

Rounding off

Rounding off a number is a way of writing the number with fewer non-***zero digits***.

Examples:
$1.02 to the nearest dollar is $1
$17.31 to the nearest dollar is $17
$0.91 to the nearest dollar is $1
308.5 to the nearest whole number is 309
0.612 to the nearest whole number is 1
0.498 to the nearest whole number is 0
13.77 to the nearest whole number is 14

When the first digit to be ignored (shown in green) is a 5, 6, 7, 8, or 9 then we 'round up'; when it is a 0, 1, 2, 3, or 4 then we 'round down'.

Examples:
2392 to the nearest hundred is 2400 (rounded up)
2342 to the nearest hundred is 2300 (rounded down)
72 849 to the nearest hundred is 72 800 (rounded down)
106 951 to the nearest hundred is 107 000 (rounded up)
687 to the nearest ten is 690 (rounded up)
9619 to the nearest thousand is 10 000 (rounded up)

Row matrix

When a ***matrix*** consist of only a row it is called a row matrix.

Example:
(4 2 8 0 −1)

S

Scalar

A scalar quantity needs only a size to describe it as opposed to a ***vector*** quantity which needs a size and a direction to describe it. Time, ***mass***, and length are examples of scalar quantities.

Scale drawing

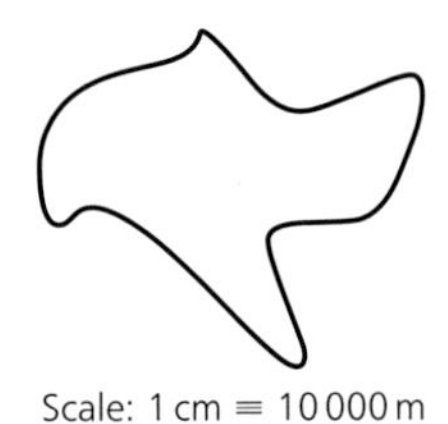

Scale: 1 cm ≡ 10 000 m

A scale drawing is a drawing of an object where the lengths on the drawing are ***directly proportional*** to the lengths on the object. The ***scale*** (***factor***) gives the relationship between the lengths on the drawing and the actual lengths.

Example: This map is a scale drawing of an island. The scale shows that 1 cm on the map represents 10 000 m on the ground.

Scale factor

When objects are ***similar***, their sizes can be compared by looking at the ***ratio*** of the lengths of their corresponding parts, e.g. the handles of the saucepans.

Linear scale factor

The linear scale factor is the number of times the length on one object is bigger than the corresponding length on the similar object.

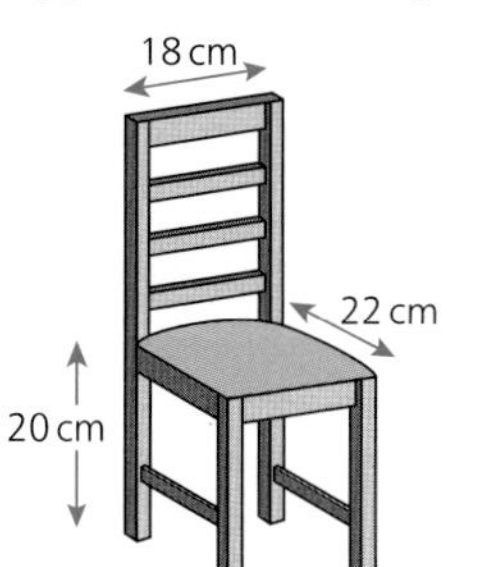

Example: The child's chair is made similar to the adult's chair on a scale 1 to 2.
The scale factor for the child's chair from the adult's is $\frac{1}{2}$.
The scale factor for the adult's chair from the child's is 2.

Scale factor of enlargement

In an ***enlargement*** a scale factor is used to produce a similar shape.

Example: The enlargement P′Q′R′ of shape PQR is made with centre O and scale factor 3.

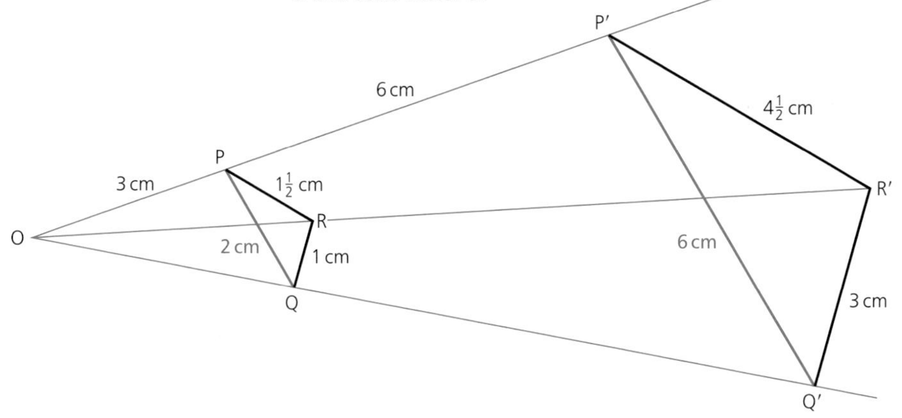

The distance of the ***image*** point is 3 times the distance of the ***corresponding object*** point from the centre of enlargement:

$OP' = 3 \times OP \qquad OQ' = 3 \times OQ \qquad OR' = 3 \times OR$

Also, the image lengths are three times the object lengths:

$P'Q' = 3 \times PQ$

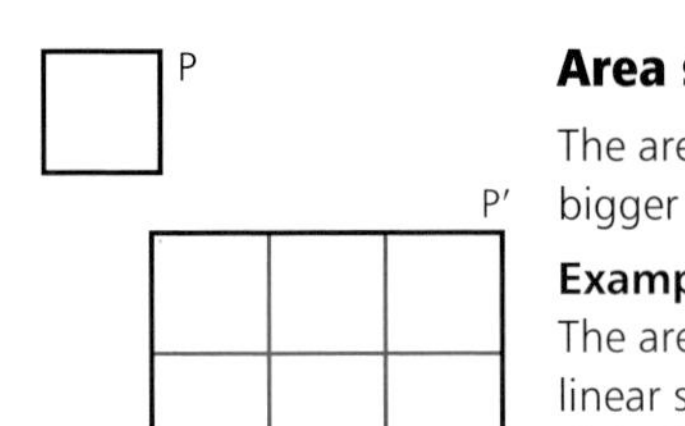

Area scale factor

The area scale factor is the number of times the ***area*** of one shape is bigger than the area of the similar shape.

Example (left):
The area P′ is 9 times the area P, so the area scale factor is 9; but the linear scale factor is 3.
If the linear scale factor is k then the area scale factor is k^2.

See **Enlargement** (Enlargement matrix)

Scalene triangle

A scalene triangle is a ***triangle*** whose sides are all of different lengths.

Examples:

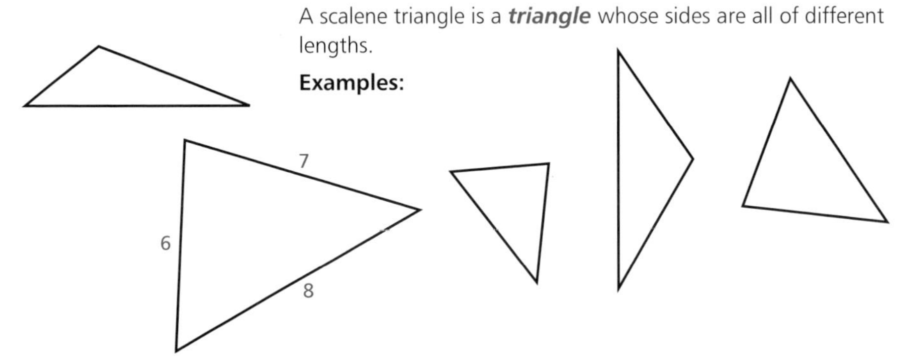

NOTE: A scalene triangle can be ***right-angled***.

Sector of a circle

A sector of a circle is a shape whose ***boundary*** is an ***arc*** of the ***circle*** and two ***radii*** of the circle. It is the shape of the top of a slice of cake or a wedge of cheese.

Examples:

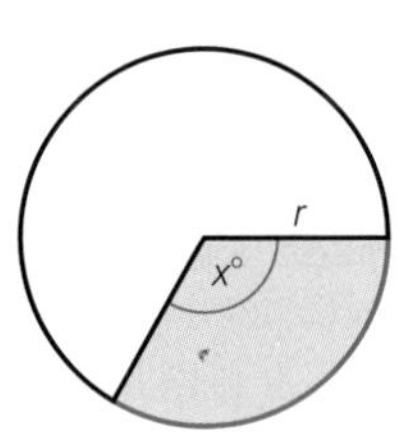

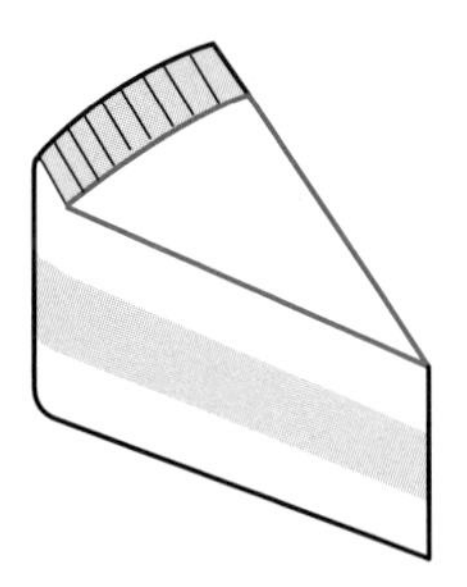

The area of a sector is $\frac{x}{360} \times \pi r^2$

Segment of a circle

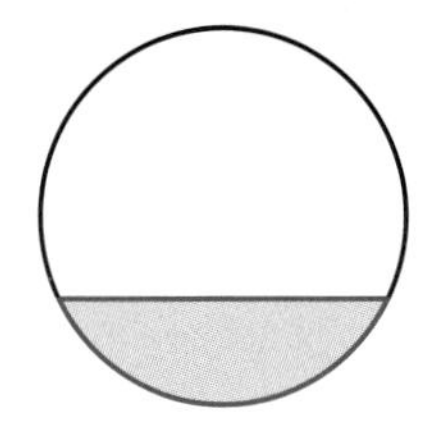

The segment of a circle is a shape whose ***boundary*** is an ***arc*** of the ***circle*** and a ***chord*** of the circle.

Examples:
A chord normally divides a circle into two segments of different sizes. The smaller segment is called the minor segment and the larger segment is called the major segment.

The ***area*** of a segment is found by subtracting the area of the ***triangle*** from the area of the ***sector***.

$\frac{x}{360} \times \pi r^2 - \frac{1}{2}r^2 \sin x$

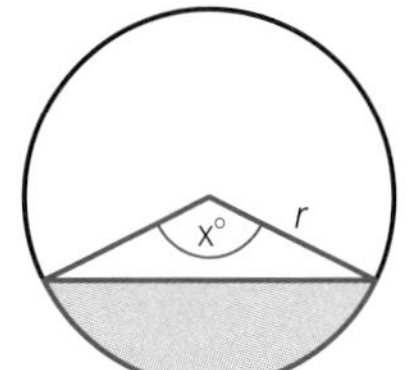

Semicircle

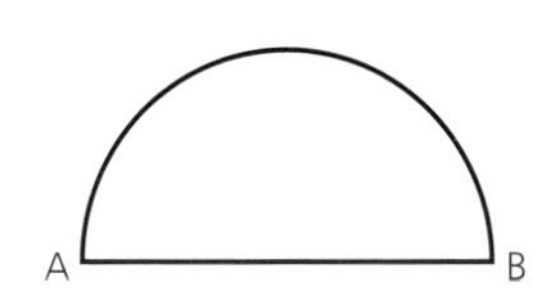

Semicircle means half a ***circle***. It is the shape formed by a ***diameter*** and the ***arc*** of the circle joining its end points.

Examples:
The ***line segment*** AB is always a diameter.

Semi-interquartile range

The semi-interquartile range is half the ***interquartile range.***

Sequence

A sequence is a set of patterns, numbers or ***expressions*** in a given order so there is a first ***term***, a second term and so on, with a rule for finding the terms.

Examples:

● ●● ●●● ... is a sequence of patterns of dots
● ●●
●

2, 4, 6, 8, ... is a sequence of ***even numbers*** starting with 2.
x, x^2, x^3, ... is a sequence of ***integer*** powers of x starting with x^1.

Series

A series is the ***sum*** of the ***terms*** of a ***sequence***.

Example: 2 + 4 + 6 + 8 + ... is a series.

Set

A set is any collection of things. The ***members*** of a set could be numbers, names, letters, shapes, ***matrices***, etc.

Examples:
The set of vowels is {a, e, i, o, u}
The set of ***whole numbers*** less than 5 is {0, 1, 2, 3, 4}
{***even numbers***} = {2, 4, 6, 8, 10, 12, 14, 16, ...}
{traffic-light colours} = {red, amber, green}

When we use some ***elements*** of a set A to form a new set, the new set is called a ***subset*** of A.

Example:
If A = {1, 3, 5, 7, 9} and B = {3, 5, 7} then
B is a subset of A.

When we put the elements of A and the elements of B into one set, this new set is called the ***union*** of A and B.

Example:
If A = {t, i, m} and B = {j, i, m} then {t, i, j, m} is the union of A and B.

NOTE: The order of the elements doesn't matter and we never put the same element in a set more than once.

When we take the elements that come in both set A and set B, then the new set is called the ***intersection*** of A and B.

Example:
If A = {Δ, □, ○} and B = {Δ, □, +, ∗} then {Δ, □} is the intersection of A and B.

See **Complement of a set, Empty set, Universal set**

Shear

A shear is a ***transformation*** in which all points slide ***parallel*** to a fixed line, or ***plane***, keeping all straight ***lines*** straight.

Examples:

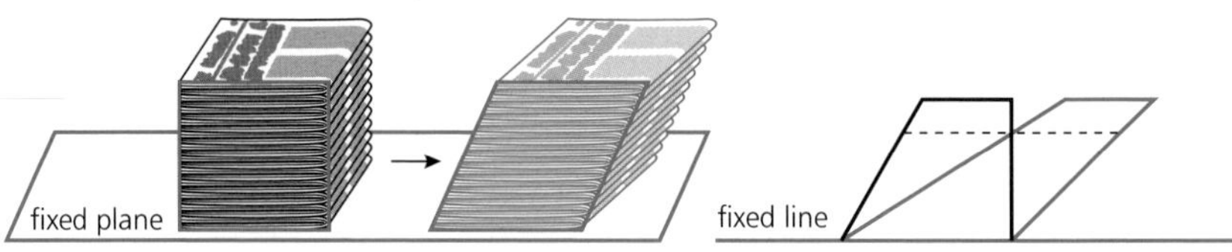

The black shape is the ***object*** and the green shape is the ***image*** after the shear.

The ***areas*** of the object shape and the image shape are the same.

Example:

The area of the black triangle is 10 and the area of the green triangle is 10.

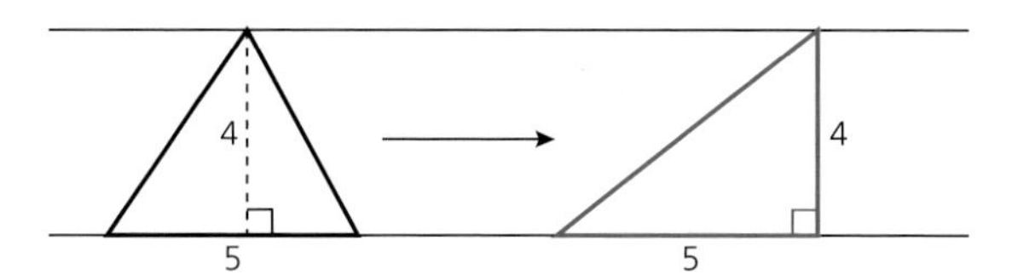

There is always an ***invariant*** line and all points move parallel to it. The invariant line need not be on the shape.

Example:

Line P is the invariant line for this shear.

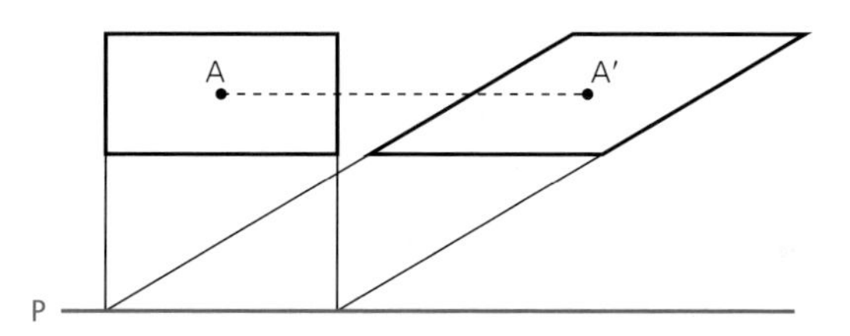

The line joining A and A′, its image, is parallel to P.

The ***ratio*** of the distance moved by points to their distance from the invariant line is ***constant***.

Shear matrix

The transformation

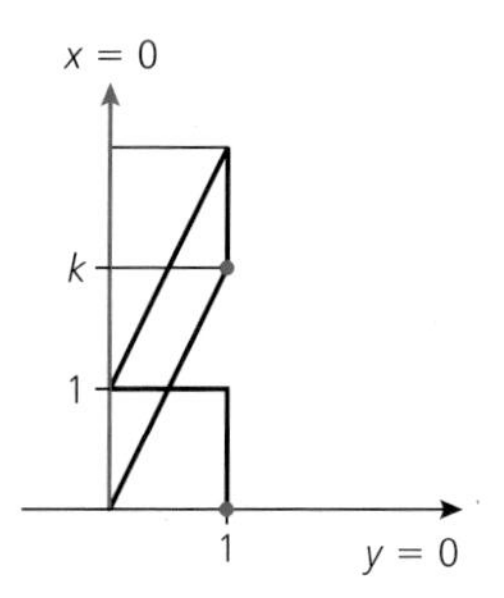

$\begin{pmatrix} x \\ y \end{pmatrix} \to \begin{pmatrix} 1 & 0 \\ k & 1 \end{pmatrix} \begin{pmatrix} x \\ y \end{pmatrix}$

represents a shear with $x = 0$ invariant and the point (1, 0) having the image (1, k)

Any ***matrix*** $\begin{pmatrix} a & b \\ c & d \end{pmatrix}$ where $a + d = 2$ and $ad - bc = 1$ will represent a shear.

Significant figure

The most significant figure (or ***digit***) in a number is the first digit (not ***zero***) that you reach, reading left to right.

Examples:
The significant figures are shown in green.

7861
54002
91 — Here, the ONE most significant figure is in green.
0.006103
0.01007
3.09988

Here, shown in green, are:

8140.09	the TWO most significant figures;
61954	the THREE most significant figures;
7089.3	the THREE most significant figures;
60.219	the FOUR most significant figures;
0.042871	the THREE most significant figures;
0.040000	the FOUR most significant figures.

When approximating to a given number of significant figures, we use ***rounding off*** on the last significant figure.

We shorten the words 'significant figures' to S.F. or sig. fig.

Similar

Two objects are similar if they are the same shape.

Examples:

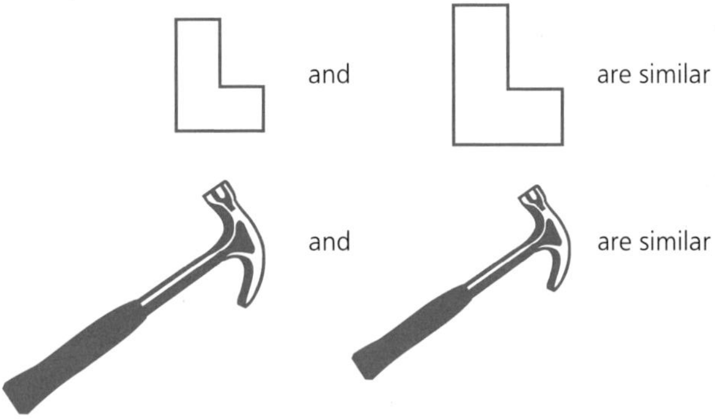

With similar objects, all the lengths of one object are a fixed number of times the corresponding lengths of the other object. This number is called the ***scale factor***.

When the scale factor is k, the ***area*** scale factor is k^2 and the ***volume*** scale factor is k^3.

Example:

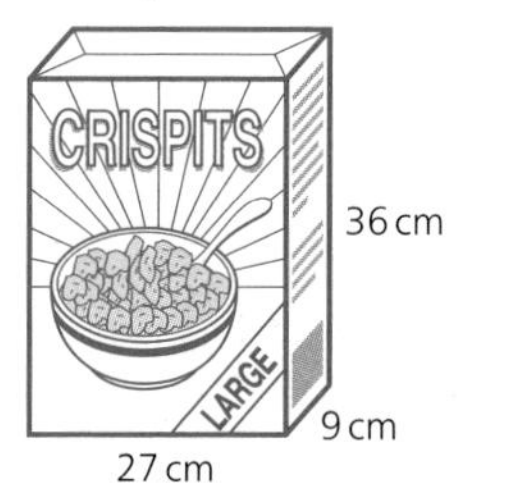

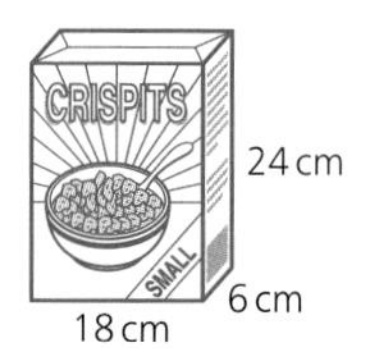

All the lengths on the large box of Crispits are $1\frac{1}{2}$ times the corresponding lengths of the small box of Crispits.

The area of the front of the large box is $\left(1\frac{1}{2}\right)^2$ times the area of the front of the small box and the volume of the large box is $\left(1\frac{1}{2}\right)^3$ times the volume of the small box.

With similar objects, corresponding ***angles*** are equal.

Example:

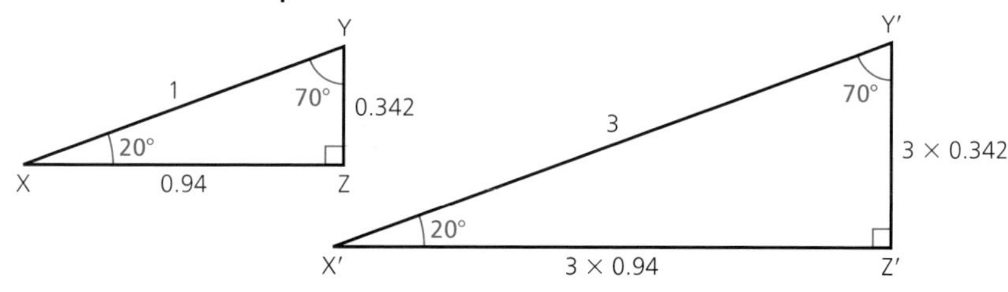

ΔXYZ and ΔX′Y′Z′ are similar

See **Enlargement**

Simple interest

Simple interest is paid only on the ***principal***.

Simultaneous equations

Simultaneous equations are ***equations*** that have the same ***solution(s)***. The equations are satisfied by the same values of the unknown quantities.

Examples:

$y = 2x$ has many solutions:

$\{\ldots (-1, -2), (0, 0),$ ***(1, 2)***, $(2, 4), (3, 6), \ldots\}$

$x + y = 3$ has many solutions:

$\{\ldots (-1, 4), (0, 3),$ ***(1, 2)***, $(2, 1), (3, 0)\ldots\}$

but if $y = 2x$ and $x + y = 3$ are taken as simultaneous equations there is only one solution, ***(1, 2)*** that is $x = 1$ and $y = 2$.

If simultaneous equations are shown ***graphically***, the solution is given by the ***intersection***(*s*) of the ***lines***.

Examples:

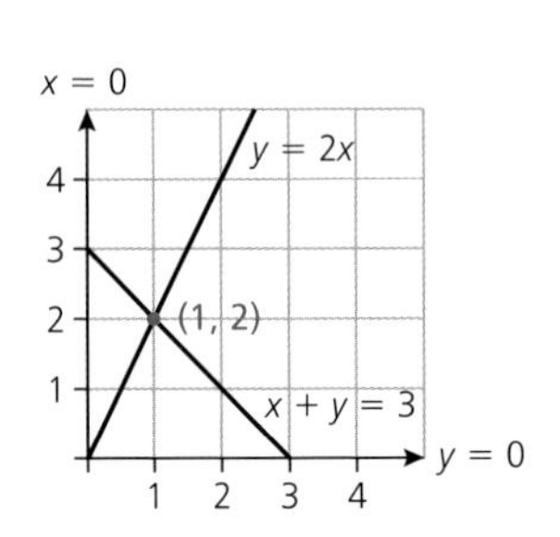

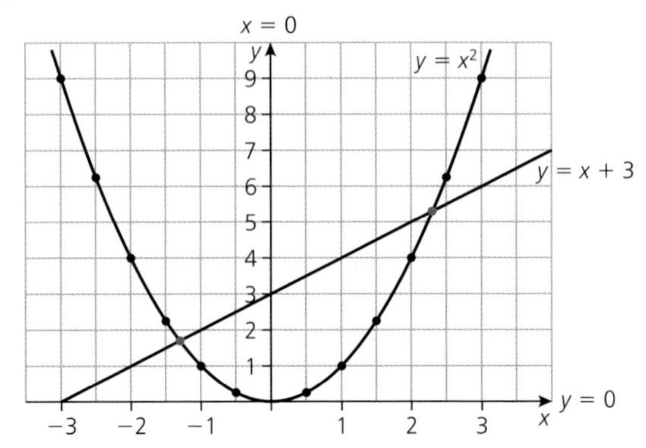

$\left.\begin{array}{l} y = 2x \\ x + y = 3 \end{array}\right\}$ have solution
$x = 1$ and $y = 2$

$\left.\begin{array}{l} y = x^2 \\ y = x + 3 \end{array}\right\}$ have solutions
$x = -1.3$ and $y = 1.7$
AND
$x = 2.3$ and $y = 5.3$

Sine

The sine of an ***angle*** θ is written as $\sin \theta$

Sine of angles less than 90°

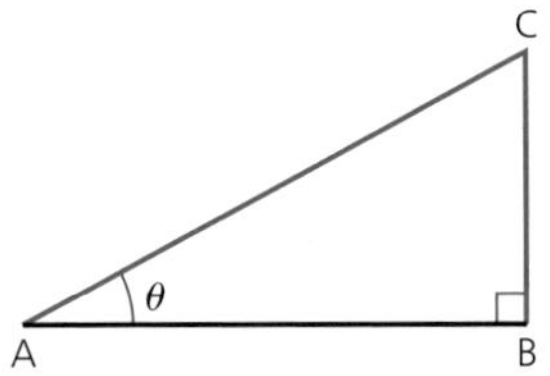

$\sin \theta = \dfrac{BC}{AC}$ $\sin \theta = \dfrac{\text{opposite side}}{\text{hypotenuse}}$

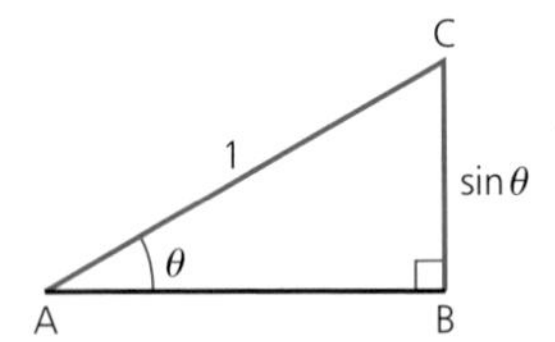

When the length of AC is 1 then
$\sin \theta = BC$

When this ***triangle*** is ***enlarged***, with ***scale factor*** r,
$QR = r \sin \theta$

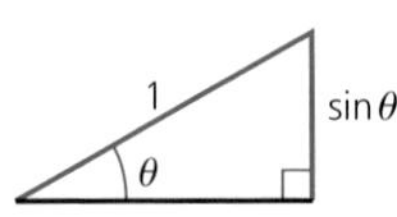

S.F. × r

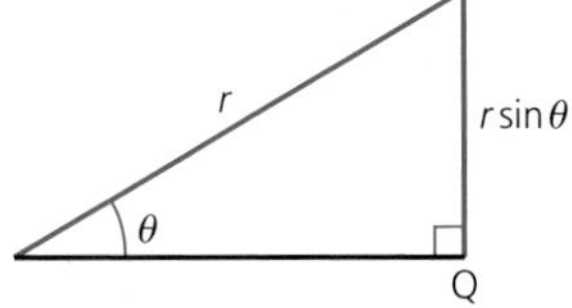

Examples:

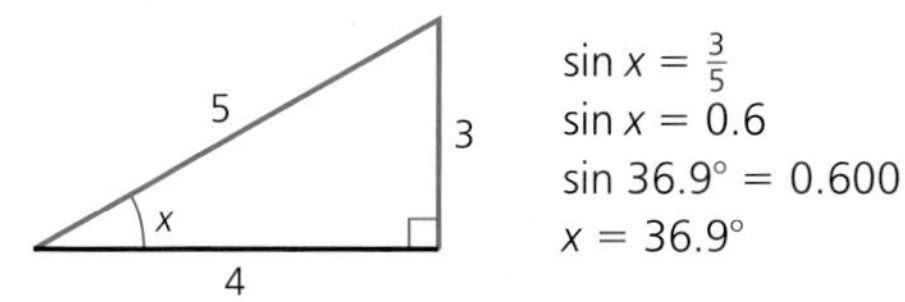

$\sin x = \frac{3}{5}$
$\sin x = 0.6$
$\sin 36.9° = 0.600$
$x = 36.9°$

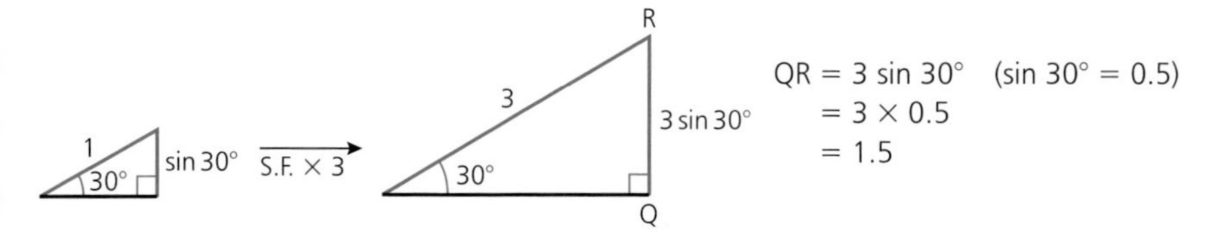

$QR = 3 \sin 30°$ $(\sin 30° = 0.5)$
$= 3 \times 0.5$
$= 1.5$

Sine of any angle

If a line OP of ***unit*** length turns through an angle θ from the *x*-***axis***; then the sine of θ is the *y*-***coordinate*** of P shown in green.

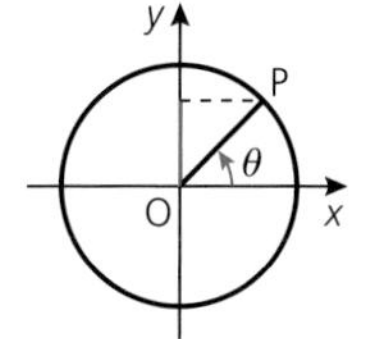

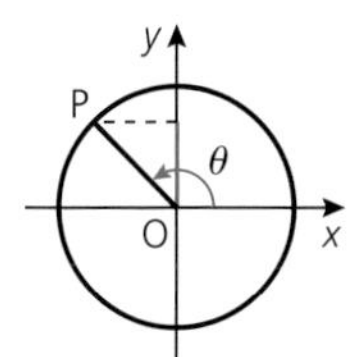

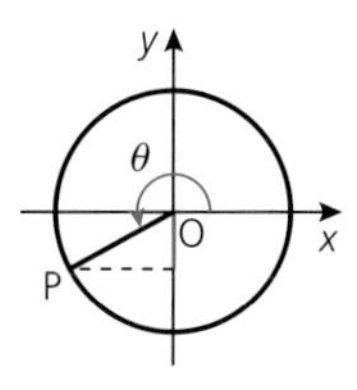

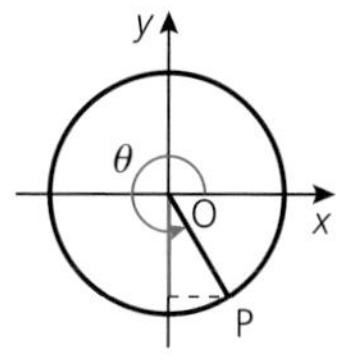

The graph of $f(x) = \sin x$ is

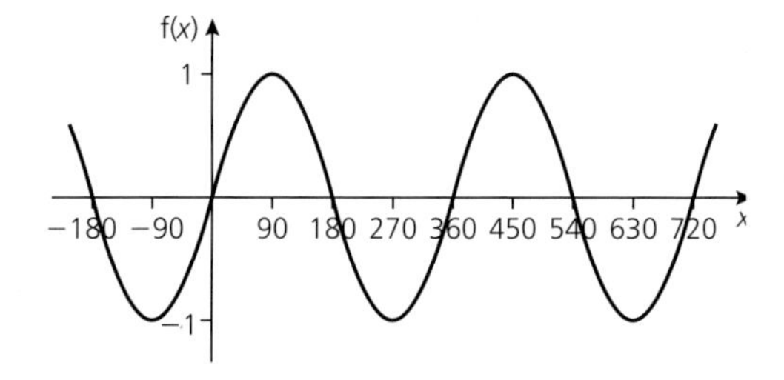

Sine formula

When two angles and the length of one side of a triangle are known, then another side can be found using the same formula:

$$\frac{a}{\sin A} = \frac{b}{\sin B} = \frac{c}{\sin C}$$

Singular matrix

A singular ***matrix*** has a ***determinant*** whose value is ***zero***.

Example: $\begin{pmatrix} 2 & 3 \\ 4 & 6 \end{pmatrix}$ is singular because $\begin{vmatrix} 2 & 3 \\ 4 & 6 \end{vmatrix} = 12 - 12 = 0$

Skew lines

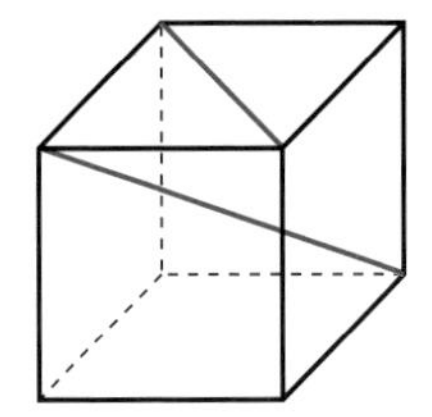

Skew lines are lines in three dimensions that are not ***parallel*** and do not ***intersect***.

Example (left): The green lines on this cube are skew lines.

Slant height

Slant height of a cone

When a ***cone*** has a ***circle*** for a base and the ***vertex*** is above the centre of the base, then the cone is a right circular cone. For such a cone, the slant height is the length of a straight ***line*** from the vertex to the circle forming the base.

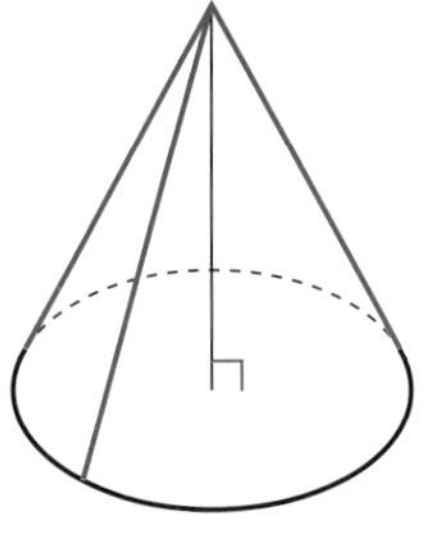

For this cone the slant height is represented by the green lines.

Slant height of a pyramid

When a ***pyramid*** has a ***regular polygon*** for a base and the ***apex*** is above the centre of the base, then the pyramid is a right pyramid. For such a pyramid, the slant height is the length of a line from the apex ***perpendicular*** to an edge of the base.

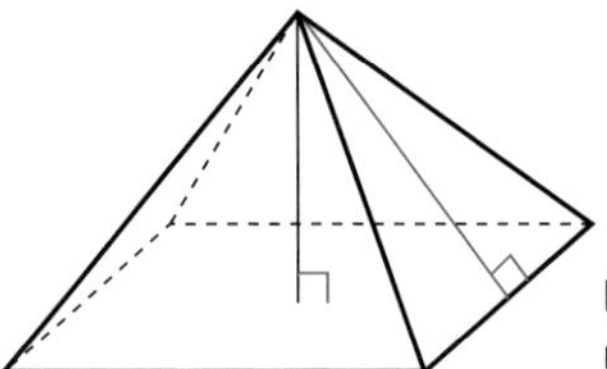

For this pyramid the slant height is represented by the green line.

Solution of an equation

The solution of an ***equation*** is the ***set*** of numbers which, when put instead of the letters, make the statement true.

Examples:

$4 + x = 11$ has solution $x = 7$ because $4 + 7 = 11$
$2 \times x = 10$ has solution $x = 5$ because $2 \times 5 = 10$
$y - 3 = 5$ has solution $y = 8$ because $8 - 3 = 5$
$x^2 = 9$ has two solutions $x = 3$ and $x = -3$
$y = x + 3$ has many solutions; one solution is $x = 2$, $y = 5$ which can be written using ***ordered pairs*** as (2, 5).
The ***solution set*** for $y = x + 3$ with x and y being ***integers*** is $\{\ldots (-1, 2), (0, 3), (1, 4), (2, 5), \ldots\}$

Solution set

When the ***solution of an equation*** or ***inequality*** is more than one number or pair, then the solution is usually called a solution set.

Examples:

$x^2 = 4$ has the solution set $\{x = +2, \text{ or } x = -2\}$
or $\{x: x = +2 \text{ or } -2\}$
For x a ***natural number***, $2x > 6$ has the solution set
$\{x > 3\}$ or $\{x: x > 3\}$
or $\{4, 5, 6, 7, 8, 9, 10, 11, \ldots\}$

Because a solution set often has many ***members*** it is helpful to show it ***graphically***.

Examples:

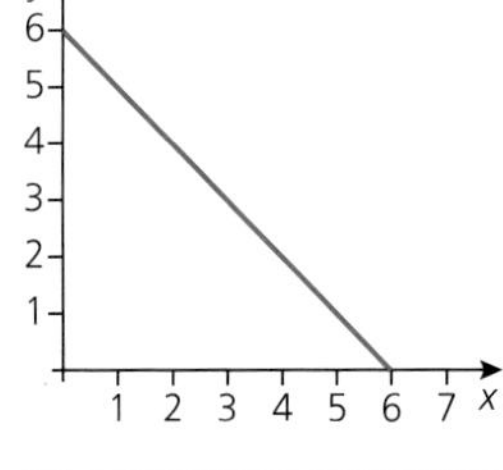

For x and y ***real***, $x + y = 6$ has the solution set shown by the green line, that is $\{(x, y): x + y = 6\}$

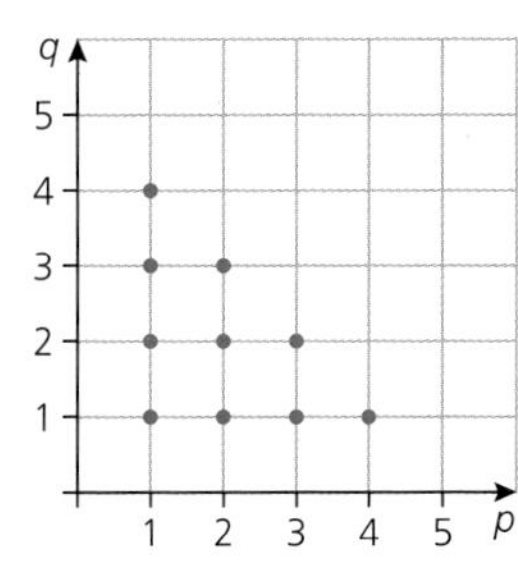

For p and q ***integers***, $p + q \leqslant 5$, $p > 0$ and $q > 0$, has the solution set shown by the green dots, that is
$\{(1, 1), (1, 2), (1, 3), (1, 4), (2, 1), (2, 2), (2, 3), (3, 1), (3, 2), (4, 1)\}$

The solution sets are shown in green:

for x real, and $x > 3$ it is

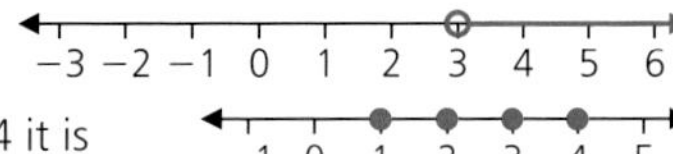

for x an integer and $0 < x \leqslant 4$ it is

for x an integer and $^-1 \leqslant x \leqslant 4$ it is

for x real and $^-2 < x \leqslant 4$ it is

The speed of a body is the distance travelled by the body per ***unit*** of time.

Examples:

A car travels along a road for 2 hours at the same speed and goes 142 ***kilometres***. The speed at which the car is travelling is

$$\frac{142 \text{ kilometres}}{2 \text{ hours}} = 71 \text{ kilometres per hour}$$
$$= 71 \text{ km/h}$$

A bullet takes 3 seconds to go 1200 metres. The speed of the bullet changes over the 3 seconds; the 'average' speed of the bullet is

$$\frac{1200 \text{ metres}}{3 \text{ seconds}} = 400 \text{ m/s}$$

For a body whose speed is changing, its speed at an instant is represented on a ***distance-time graph*** by the ***gradient*** of the curve at that instant.

Example: A marble rolls down a slope and the distance travelled is shown by this graph.

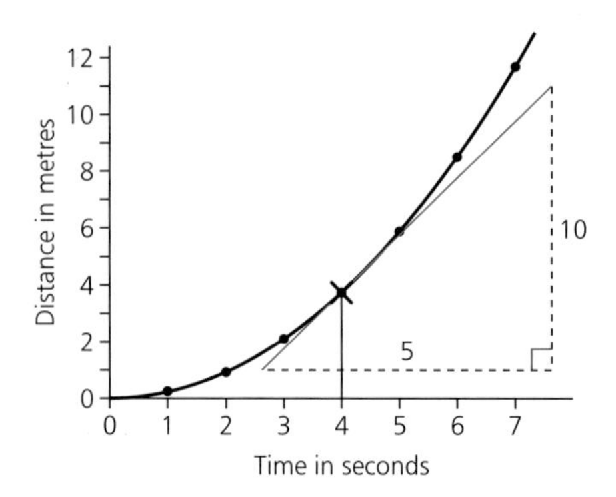

The speed of the marble after 4 seconds is given by the gradient of the green line:

$$\frac{10 \text{ metres}}{5 \text{ seconds}} = 2 \text{ m/s}$$

See **Acceleration, Velocity**

Speed-time graph

A speed-time ***graph*** shows a journey by plotting the ***speed*** against the time taken.

Example: This speed-time graph illustrates a boy's walk from home to school.

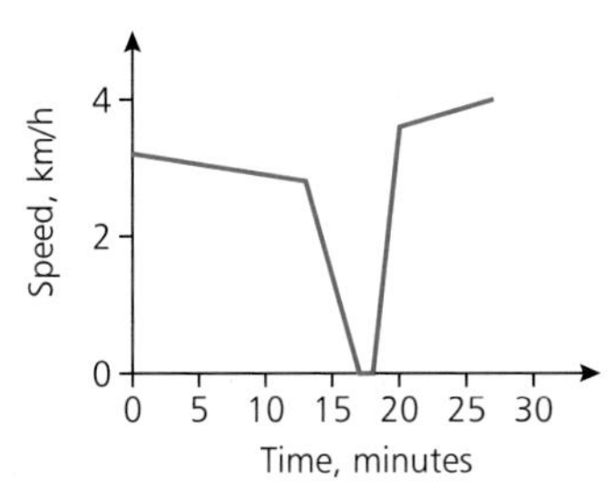

See and compare with **Distance–time graph**

Sphere

A sphere is the mathematical name for a perfectly round ball.

Examples:

A soap bubble blown into the air is very nearly a sphere.

The Earth and the other planets are roughly spheres.

A sphere is the ***set*** of points in space which are all the same distance, called the ***radius***, from a fixed point called the centre.

If the radius is r then the surface ***area*** of the sphere is given by $A = 4\pi r^2$ and the ***volume*** of the solid formed by the sphere is given by $V = \frac{4}{3}\pi r^3$

Square

A square is a ***rectangle*** with four equal sides.
A square is a ***regular quadrilateral***.

Examples:

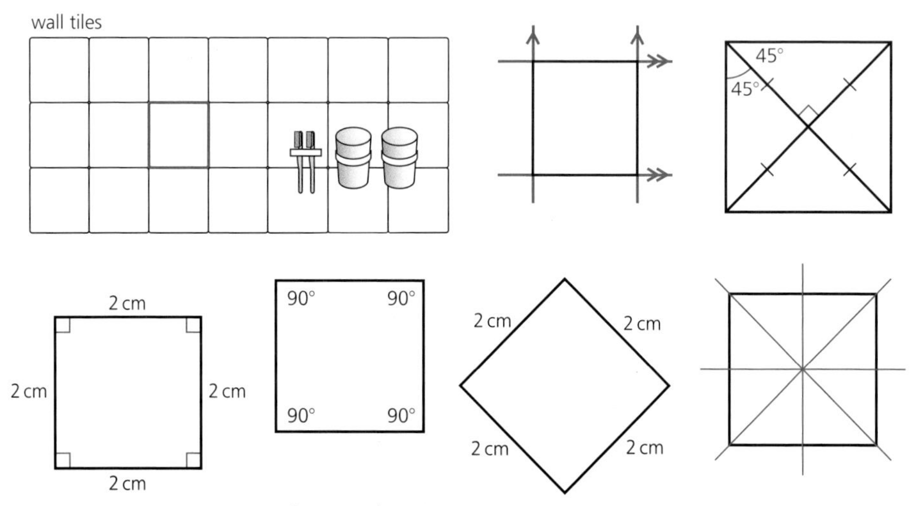

A square has:

a) four ***lines of symmetry***;
b) ***rotational symmetry*** of ***order*** 4;
c) opposite sides ***parallel***;
d) all four ***angles*** of the figure equal to 90°;
e) ***diagonals*** equal and ***bisecting*** each other at ***right-angles***.

Square centimetre

A square centimetre is a ***unit*** for measuring ***area*** and is equal to the area of a ***square*** whose sides are one ***centimetre*** in length. Square centimetre is shortened to cm^2.

Examples:

This area 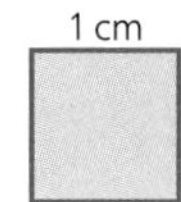

is 1 cm^2

9 cm
A
6 cm

The area of A is 6 cm × 9 cm = 54 cm^2

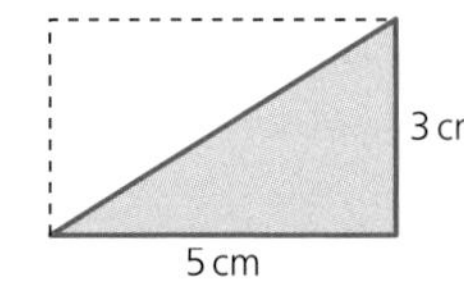

The area of the ***triangle*** is $\frac{1}{2}$ of 3 × 5 cm^2
$= \frac{1}{2} \times 15$ cm^2
= 7.5 cm^2

Square metre

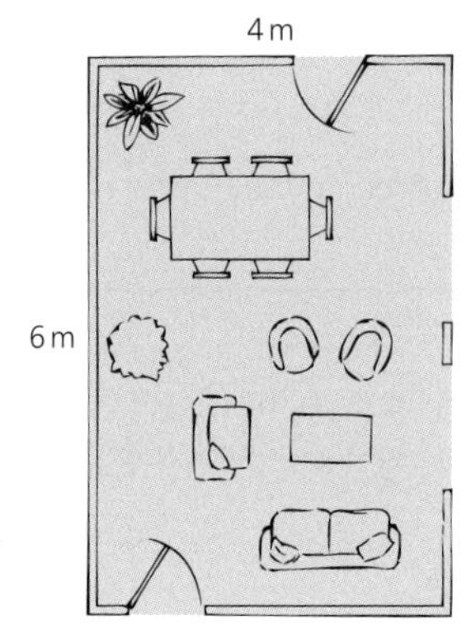

A square metre is the standard ***unit*** for measuring ***area***. One square metre is the same area as a ***square*** whose sides are one ***metre*** in length. Square metre is shortened to m^2.

Examples:

This area is 1 m^2

The area of this room is
$4 \times 6\ m^2 = 24\ m^2$

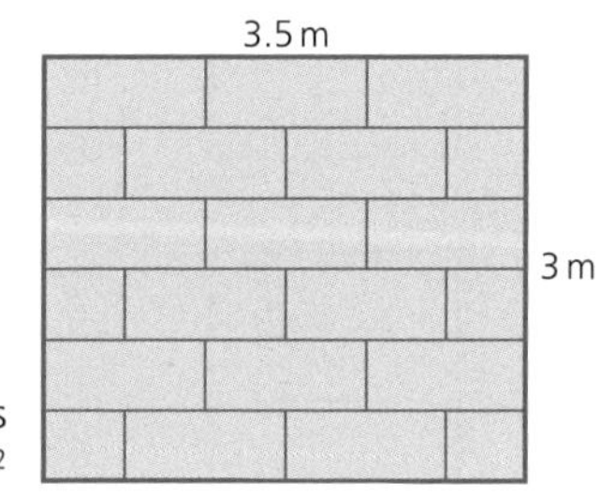

The area of this wall is
$3.5 \times 3\ m^2 = 10.5\ m^2$

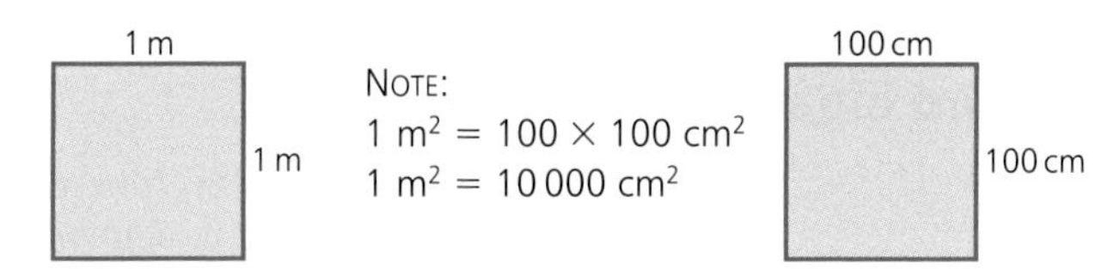

NOTE:
$1\ m^2 = 100 \times 100\ cm^2$
$1\ m^2 = 10\,000\ cm^2$

Square number

A square number is a number that can be shown as a pattern of dots in the shape of a ***square***.

Examples:

$4 = 2 \times 2$ $16 = 4 \times 4$ $25 = 5 \times 5$

```
• •        • • • •        • • • • •
• •        • • • •        • • • • •
           • • • •        • • • • •
           • • • •        • • • • •
                          • • • • •
```

{square numbers} = {1, 4, 9, 16, 25, 36, 49, 64, 81, 100, …}

NOTE: The number 1 is a square number but not a ***rectangle number***.

Square of a number

The square of a number is that number multiplied by itself. We often say it is the number 'squared'.

Example: 6 squared is 36
A number squared means the same as the number to the ***power*** of 2.

Square root

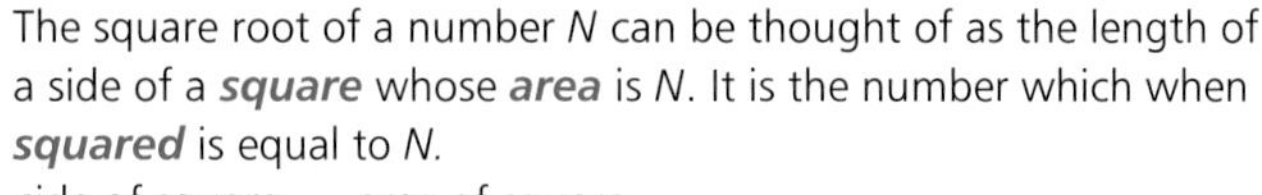

The square root of a number N can be thought of as the length of a side of a ***square*** whose ***area*** is N. It is the number which when ***squared*** is equal to N.

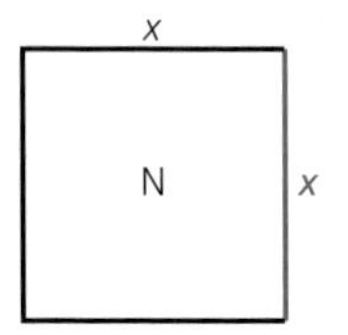

side of square   area of square
x N

x is the square root of N which is written

$x = \sqrt{N}$

Examples:

side	area
3	9

$3 = \sqrt{9}$

3
9
3

$\sqrt{36} = 6 \qquad \sqrt{49} = 7 \qquad \sqrt{1} = 1 \qquad \sqrt{0.25} = 0.5$

NOTE: The ***equation*** $x^2 = 4$ has two ***solutions*** $x = {}^{+}\sqrt{4}$ and $x = {}^{-}\sqrt{4}$ that is

$x = +2$ and $x = -2$

Standard index form

A number is in standard (***index***) ***form*** when it is written as a number between 1 and 10 multiplied by a ***power*** of 10.

Examples:

$$400\,000 = 4 \times 10^5$$
$$9\,650\,000 = 9.65 \times 10^6$$
$$1013 = 1.013 \times 10^3$$
$$0.00308 = 3.08 \times 10^{-3}$$
$$7.894 = 7.894 \times 10^0 = 7.894 \times 1 = 7.894$$

Standard deviation

Standard deviation is a measure of the spread of ***data***. It is also called the root mean square deviation.

The ***formula*** for finding the standard deviation is

$$\sqrt{\frac{\sum_{i=1}^{n}(\bar{x} - x_i)^2}{n}}$$ where $\bar{x}$ is the mean value.

Stationary point

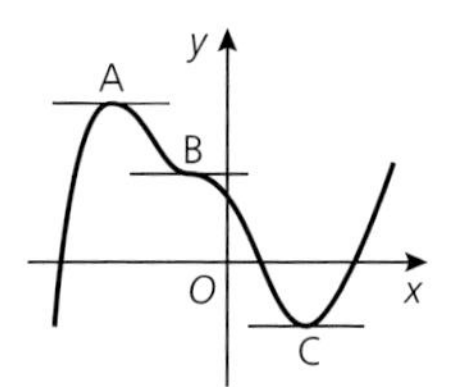

A stationary point on the ***graph*** of $y = f(x)$ is a point on the graph where the ***gradient*** of the curve is ***zero***, i.e. $\frac{dy}{dx} = 0$.

Maximum values and ***minimum values*** are stationary points.

Example: A, B and C are stationary points on this curve.

Statistic

Statistics is the name given to the collection and analysis of ***data*** which can be numerical or descriptive.

A statistic is a single value worked out using data, for example, the ***mean*** or ***standard deviation***.

Stretch

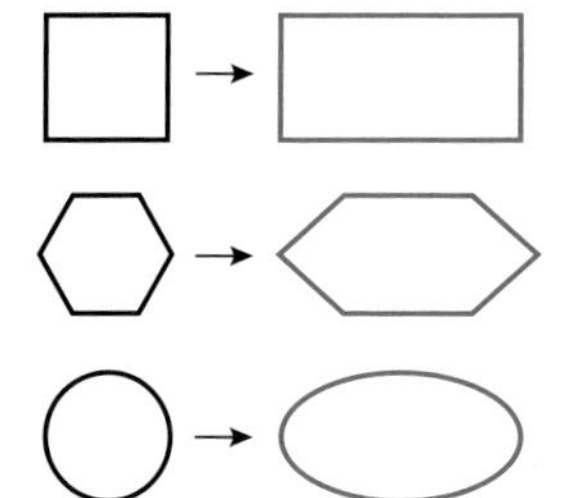

One-way stretch

A stretch is a ***transformation*** in which shapes are pulled or stretched in one direction.

Examples (left):
The green shapes show the ***images*** of the black shapes after a stretch across the page.

There is an ***invariant line***, and all other points move on lines ***perpendicular*** to this line.

Example:

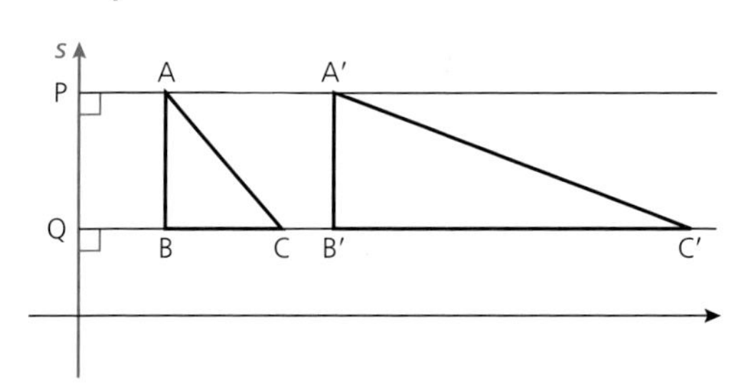

The green line is invariant and A′B′C′ is the image after a stretch in the direction of the *x-**axis***.

Lines PAA′ and QBB′ are perpendicular to the invariant line.

There is a linear ***scale factor*** k which gives the amount of stretch. For any point A, with line PA perpendicular to the invariant line s, its image A′ is given by

$PA' = k \times PA$

Example: The invariant line is $x = 0$. The image after a stretch with scale factor 3 is shown in green.

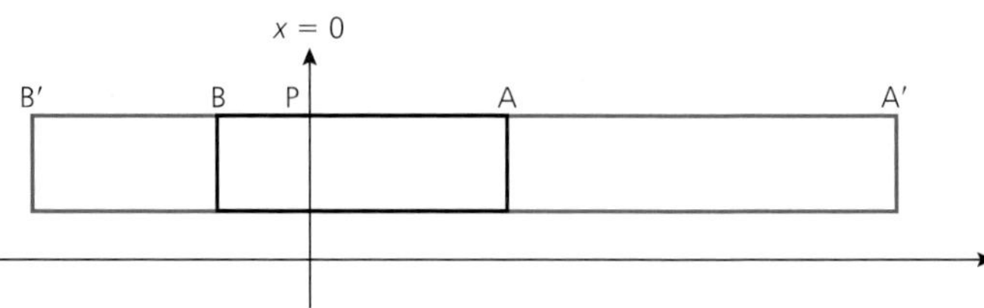

Stretch matrix

The transformation $\begin{pmatrix} x \\ y \end{pmatrix} \rightarrow \begin{pmatrix} 1 & 0 \\ 0 & k \end{pmatrix} \begin{pmatrix} x \\ y \end{pmatrix}$

represents a stretch with $y = 0$ and scale factor k.

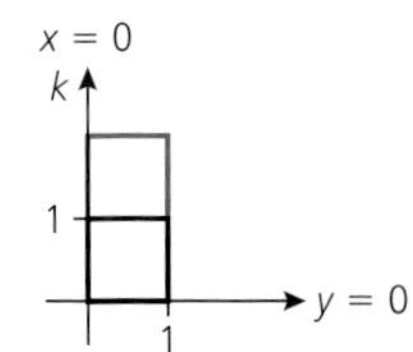

Two-way stretch

A two-way stretch is the single transformation representing stretches with two different invariant lines.

Example:

The image A′B′C′D′ shows the two-way stretch on ABCD:
$x = 0$ invariant and scale factor 2,
$y = 0$ invariant and scale factor 3.

Structure

Many ***sets*** of ***elements*** have ***operations*** on them which show similar properties.

For an operation $*$ on a set, some of the properties we ask for are:

a) Is the operation $*$ on the set ***closed***?
b) Is there an ***identity*** element in the set?
c) If there is an identity element, does each element of the set have an ***inverse*** in the set?
d) Does the operation $*$ on the set show that it is
   i) ***associative***?
   ii) ***commutative***?
   and, when there is a second operation $\circ$ on the same set, that
   iii) $*$ is ***distributive*** over $\circ$?

A set with an operation which has the four properties of closure, identity, inverse, and associativity is called a group.

Examples:

The following sets with operations are each a group:

Numbers under addition;
Numbers (omitting ***zero***) under multiplication;
Vectors under addition;
The set of all 2 by 2 ***matrices*** (with ***determinant*** not zero) under matrix multiplication.

Subject of a formula

The subject of a formula is a single letter on one side of the equal sign that does not appear on the other side.

Example: A is the subject of the ***formula*** $A = \frac{1}{2}\pi r^2 h$.

Subset

If every ***element*** of ***set*** B is also an element of the set A then B is a subset of A.

Example: If A is {a, b, c, d} then all the possible subsets B of set A are:
{a} {b} {c} {d} {a, b} {a, c} {a, d}
{b, c} {b, d} {c, d} {b, c, d} {a, c, d}
{a, b, d} {a, b, c} {a, b, c, d} and { }
When B is a subset of A we write $B \subset A$

NOTE: In the example above, all the subsets except { } and {a, b, c, d} are PROPER SUBSETS of A.

Subtend an angle

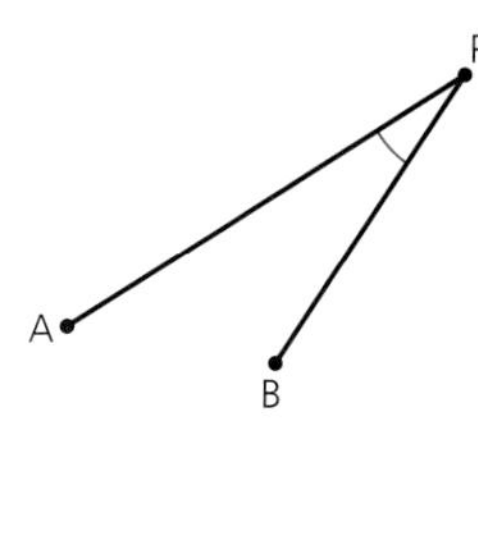

The angle subtended by two points A and B at a point P is the ***angle*** between the ***lines*** AP and BP.

Examples:

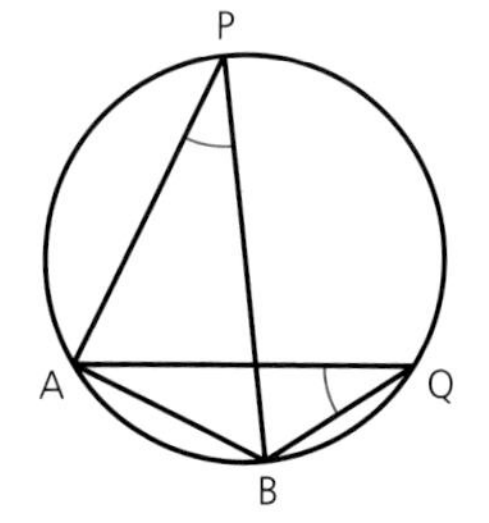

The angle subtended by a ***chord*** AB of a ***circle*** at a point P on its ***circumference*** is equal to the angle it subtends at Q.

The angle subtended by the end points C, D of a ***diameter*** of a circle at a point T on its circumference is a ***right-angle***.

Subtract

Subtract is the mathematical word for 'take away'.

Sum

Sum is the mathematical word for 'add up'.

Supplementary angles

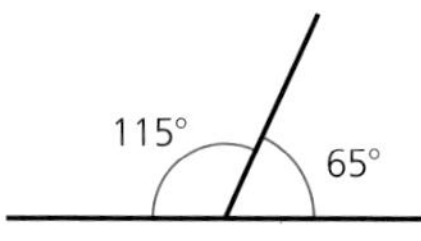

If two ***angles***, such as 115° and 65°, add up to 180° they are called supplementary angles.

$115° + 65° = 180°$.
115° and 65° are supplementary angles.

The opposite angles of a ***cyclic quadrilateral*** are always supplementary.
$x + y = 180°$

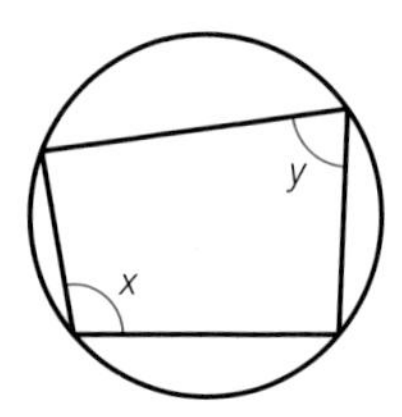

Surd

A surd is an ***irrational*** root of a number.

Example: $\sqrt{2}$, $\sqrt{3}$, $\sqrt{7} + 1$, $9 - \sqrt{3}$, $\sqrt{8}$, $\sqrt[3]{16}$

Surface area

The surface area of a solid is the total ***area*** of all the ***faces*** and all external curved surfaces of the solid.

Example: The surface area of this cylinder is the sum of the areas of the top and bottom circles and the curved surface.

Symmetry

Symmetry is the kind of pattern that a shape has.

Symmetry of plane shapes

There are two kinds of symmetry that a ***plane shape*** can have:

a) ***line symmetry***;
This shape has one line of symmetry shown in green;

b) ***rotational symmetry***;
This shape has rotational symmetry of ***order*** 3.

Symmetry of solid shapes

A solid can have:

a) ***plane symmetry***;
This fish has one plane of symmetry shown in green;

b) ***rotational symmetry***;
This propeller has rotational symmetry of order four.

Tangent of an angle

The tangent of an ***angle*** θ is written as tan θ.

Tangent of angles less than 90°

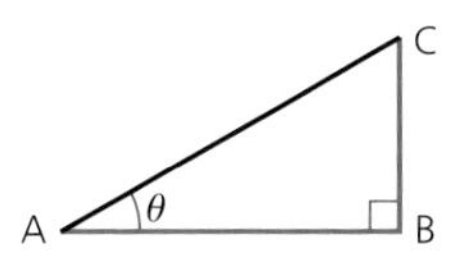

$\tan \theta = \frac{BC}{AB}$ $\tan \theta = \frac{\text{opposite side}}{\text{adjacent side}}$

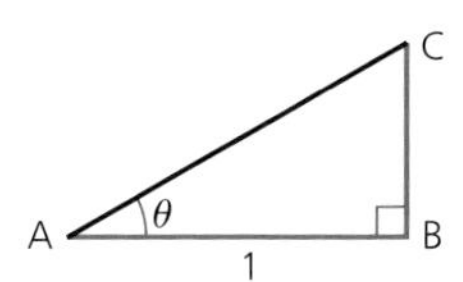

When the length of AB is 1 then $\tan \theta = BC$

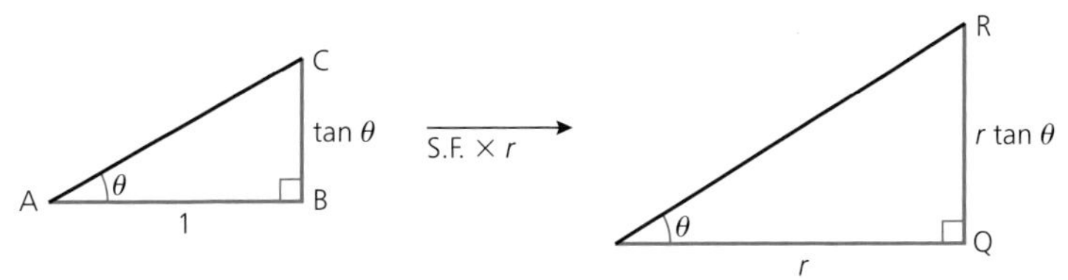

When this ***triangle*** is ***enlarged***, with ***scale factor*** r, $QR = r \tan \theta$

Examples:

$\tan x = \frac{3}{4}$
$\tan x = 0.75$
$\tan 36.9° = 0.751$
$x = 36.9°$

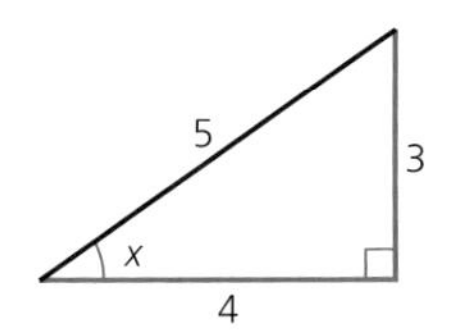

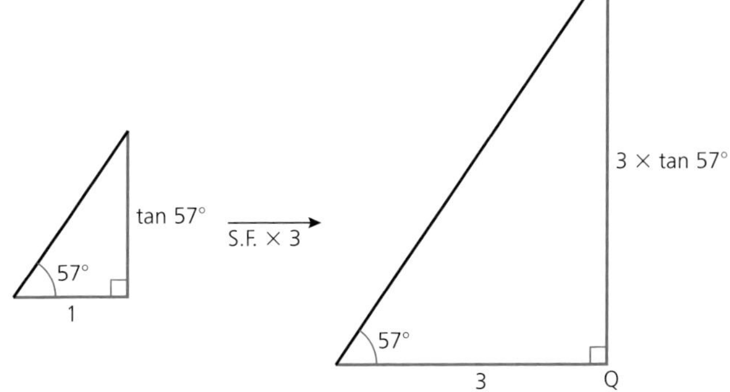

$QR = 3 \times \tan 57°$
$= 3 \times 1.54$
$= 4.62$

Tetrahedron

A tetrahedron is a solid with four ***triangular faces***. It is the same as a triangular-based ***pyramid***.

Examples:

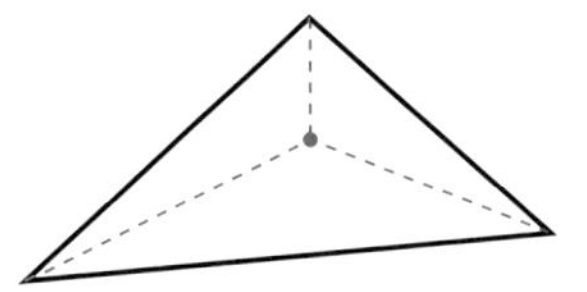

This is an irregular tetrahedron; it has four triangular faces but the triangles are not all the same shape and size

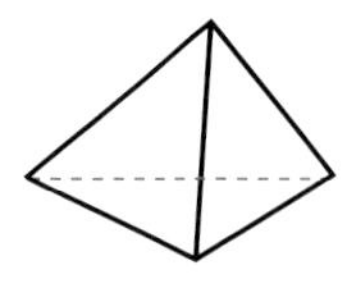

This is a ***regular*** tetrahedron; all four faces are ***equilateral triangles***.

Tonne

A tonne is a ***unit*** of ***mass***. One tonne is equal to 1000 ***kilograms***. We shorten tonne to t.

Example (left): This car has a mass of about one tonne.

Transformation

When we are ***mapping*** points and ***lines***, rather than numbers, we use the word transformation.

A transformation describes the ***relation*** between any point and its ***image*** point.

The diagram shows the transformations ***rotation***, ***reflection***, ***translation*** and ***enlargement***. A ***matrix*** can be used to describe a transformation. For a translation this is a column matrix or ***vector***.

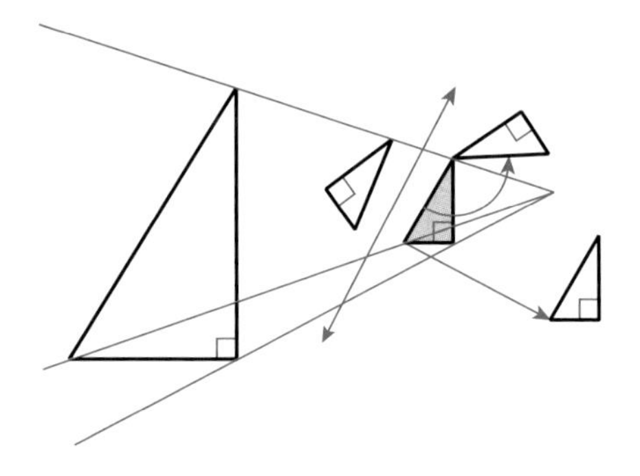

Translation

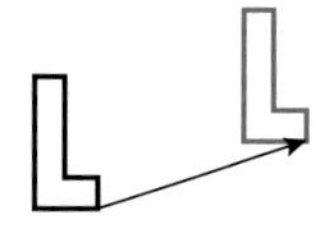

A translation is a ***transformation*** in which a shape slides without turning. Every point moves the same distance and in the same direction.

Examples:

In each case the green shape is the ***image*** after a translation.

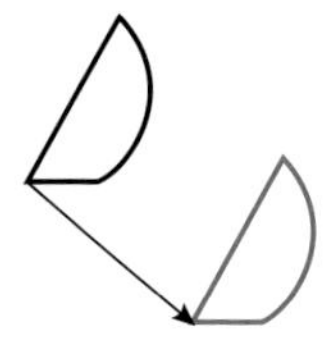

A translation can be described by a ***vector***.

Examples:

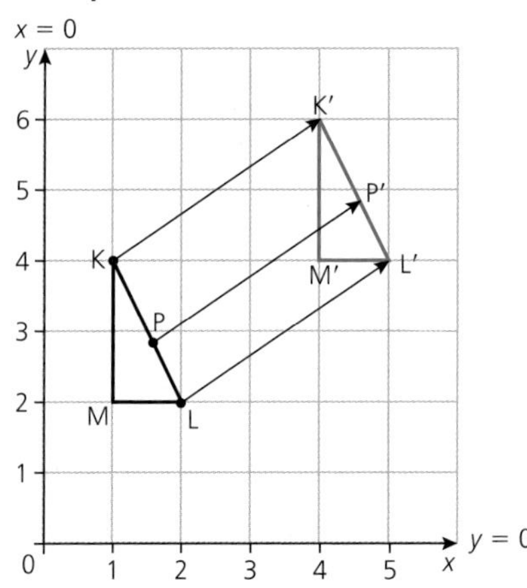

The green triangle shows the image of triangle KLM after a translation described by the vector $\begin{pmatrix} 3 \\ 2 \end{pmatrix}$. K(1, 4) maps onto K′(4, 6).

L(2, 2) is mapped onto L′(5, 4) by the translation $\begin{pmatrix} 3 \\ 2 \end{pmatrix}$.

If P is (x, y) its image P′ is found by $\begin{pmatrix} x \\ y \end{pmatrix} \rightarrow \begin{pmatrix} x + 3 \\ y + 2 \end{pmatrix}$

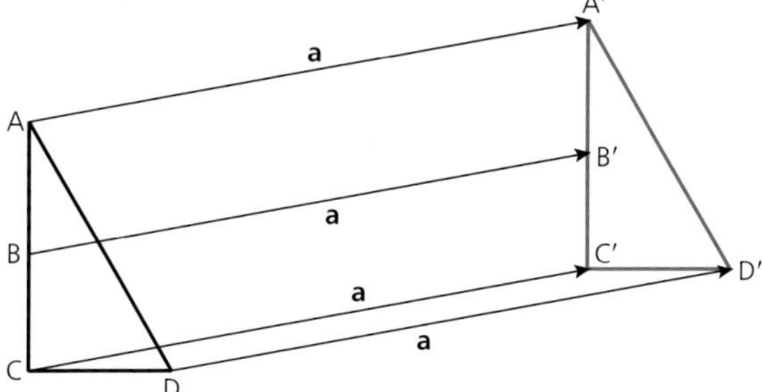

The image A′B′C′D′ is shown in green for a translation of ABCD. Each point has moved the same distance and direction, described by vector **a**.

The vector **a** can be represented by this ***line***.

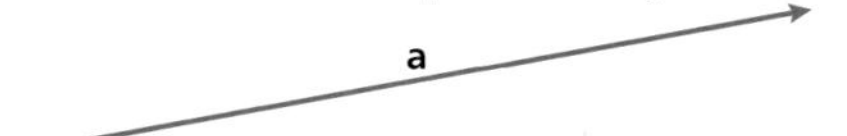

Translational symmetry

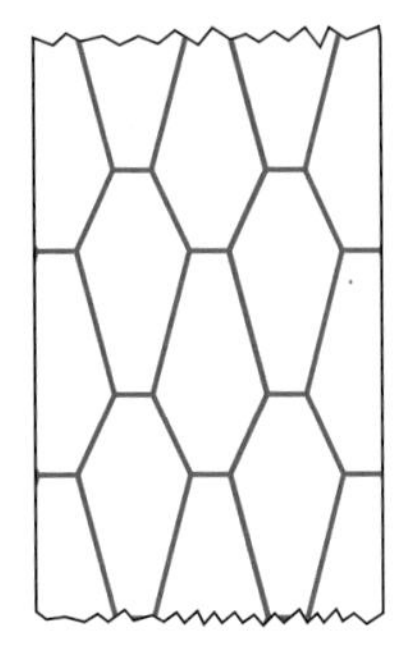

A shape has translational symmetry if it looks the same after a ***translation***. This means that the shape can be divided by ***lines*** into shapes that look the same.

Example: Material with a design on it usually has translational symmetry because it has a repeating pattern.

Transpose

The transpose of a ***matrix*** **A** is the matrix **A′** made by changing over the rows and columns.

Example:

If $\mathbf{A} = \begin{pmatrix} 2 & 0 \\ 1 & 2 \\ 1 & 0 \end{pmatrix}$ then $\mathbf{A}' = \begin{pmatrix} 2 & 1 & 1 \\ 0 & 2 & 0 \end{pmatrix}$

If the matrix **A** represents a ***relation***, then the matrix **A′** represents the inverse relation.

Transversal

A transversal is a straight ***line*** that crosses a set of ***parallel*** lines.

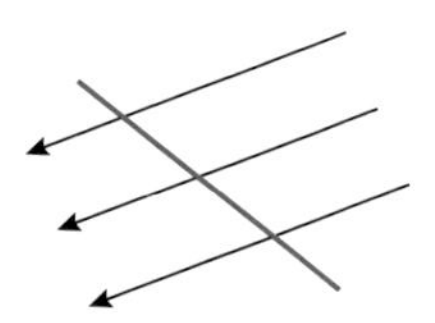

Trapezium

A trapezium is a ***quadrilateral*** with ONE pair of sides ***parallel***.
Examples:

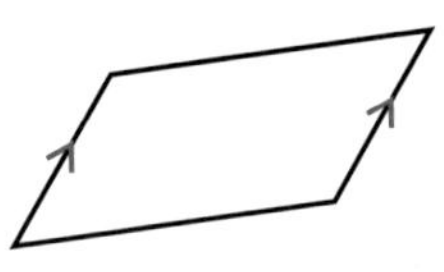

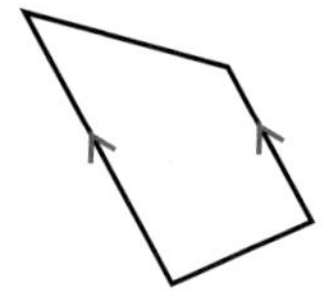

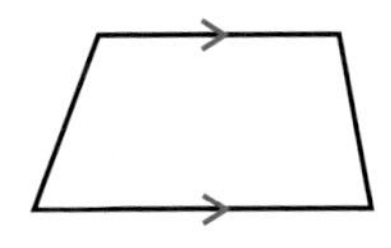

The ***area*** of a trapezium with parallel sides of length a and b is given by
Area of trapezium $= \frac{1}{2}(a + b) \times$ height

Trapezium rule

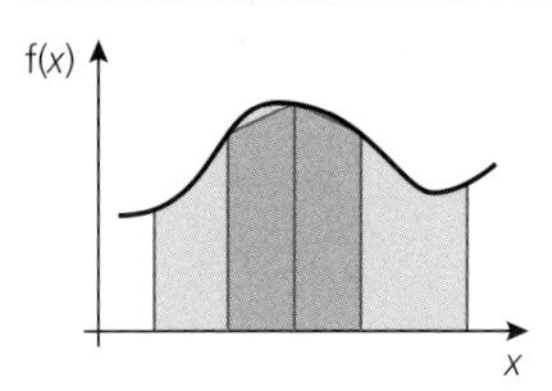

The trapezium rule is a method for ***estimating*** the ***area*** enclosed by a curve, the *x*-***axis*** and two values of *x*.

The area is divided into equal width strips by ***lines parallel*** to the *y*-axis which are joined where they cross the curve to form ***trapeziums***. The total area is approximately the sum of the areas of these trapeziums.

Triangle

See **Equilateral triangle, Isosceles triangle, Right-angled triangle, Scalene triangle**

A triangle is a ***plane shape*** bounded by three straight ***lines***.

Examples:

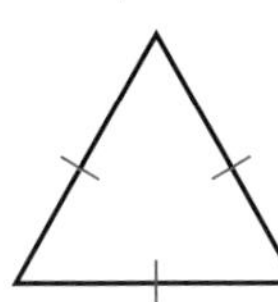
equilateral triangle

isosceles triangle

right-angled triangle

scalene triangle

Triangle law

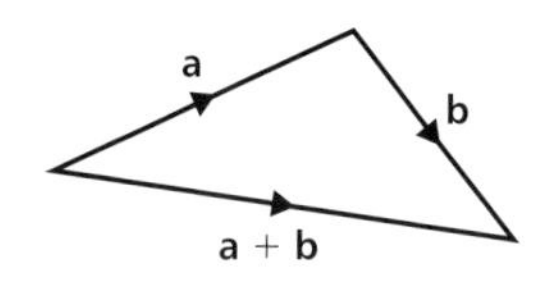

Two ***vectors*** are added using the triangle law.
The two vectors form two sides of the ***triangle*** going in the same sense. The vector represented by the third side of the triangle and going in the opposite sense to the other two vectors represents the ***sum*** or resultant of the two vectors.

Triangle number

A triangle number is a number that can be shown as a pattern of dots in the shape of a ***triangle***.

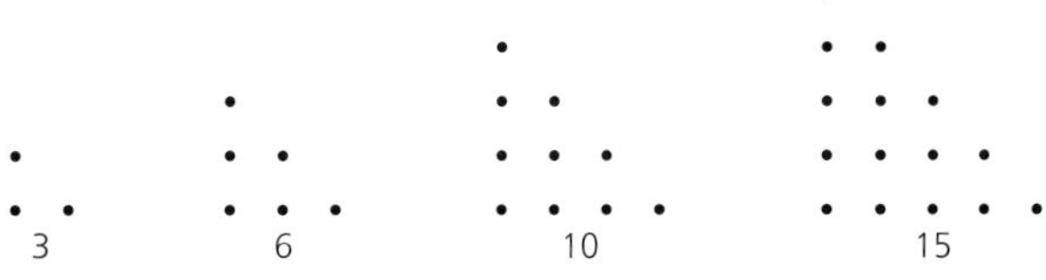

{triangle numbers} = {1, 3, 6, 10, 15, 21, 28, 36, 45, ...}

NOTE:

$10 = 1 + 2 + 3 + 4$

$21 = 1 + 2 + 3 + 4 + 5 + 6$

$36 = 1 + 2 + 3 + 4 + 5 + 6 + 7 + 8$

U

Union

The union of two ***sets*** A and B is the set of ***elements*** that are in A, or in B, or are in both A and B.

Example:

If A = {1, 4, 7, 11, 14}
and B = {2, 4, 10, 14}
then A ∪ B = {1, 2, 4, 7, 10, 11, 14}

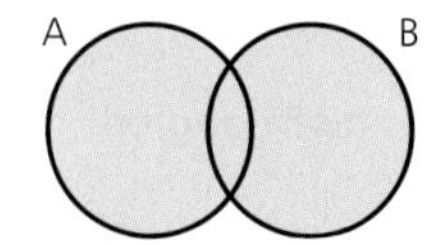

The union of sets is shown by the ***symbol*** ∪

On a ***Venn diagram*** the union of set A and set B is shown by the set of points shaded green.

A ∪ B is shaded green.

Unit vector

A unit vector has a magnitude of 1 ***unit***.

Units of measurement

The standard unit for measuring length is a ***metre***. Two smaller units of length are ***millimetre*** and ***centimetre***. A larger unit of length than a metre is a ***kilometre***.
The units for ***area*** are usually ***square centimetre*** or ***square metre***. Another unit sometimes used for large areas is ***hectare***. One hectare is 10 000 m^2.
The units for ***volume*** are usually ***cubic centimetre*** or ***cubic metre***. Another unit sometimes used for liquids or gases is ***litre***.
The standard unit for measuring ***mass*** is a ***kilogram***. A smaller unit of mass is ***gram***. A larger unit that is sometimes used is ***tonne***.

Universal set

The universal ***set*** is the set of ALL ***elements*** being considered. We use the symbol U to represent the universal set.

Example:

If U = {***natural numbers*** less than 10}
and A = {***prime numbers***}
then A′ = {1, 4, 6, 8, 9}

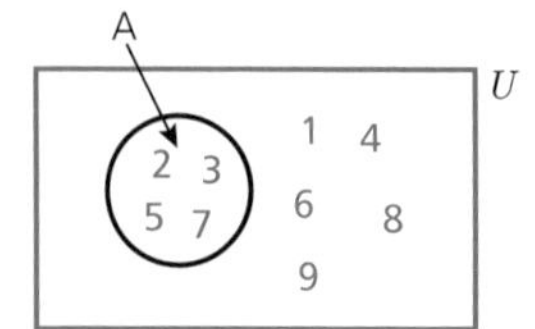

On a ***Venn diagram*** the universal set is shown by a ***rectangle***.

Example:

U = {natural numbers less than 10}
A = {prime numbers}

See **Complement of a set**

Variable

When letters such as x and y are used in ***formulas*** and mathematical ***expressions*** to stand for different numbers they are called variables, in contrast to letters which have a fixed value and are called ***constants***.

Examples:

In the formula for the area for a ***circle***, $A = \pi r^2$, r and A are variables, but π is a constant.

In the equation for a straight line, such as $y = 3x + 2$, x and y are variables, but 2 and 3 are constants.

See **Coefficient**

Variance

Variance is the square of the ***standard deviation***.

Vector

A vector is a ***line segment*** in a given direction. It is usually shown as a column of numbers like this, when it is called a ***column vector***.

$$\begin{pmatrix} 2 \\ -3 \\ 6 \end{pmatrix} \begin{pmatrix} 1 \\ 7 \end{pmatrix}$$

We often describe a vector by using a letter. The letter must be either underlined with a 'squiggle' like this $\underset{\sim}{a}$ or printed in 'bold type' like this **a**.

Examples:

$\underset{\sim}{p} = \begin{pmatrix} 3 \\ 2 \\ -5 \end{pmatrix}$ $\mathbf{b} = \begin{pmatrix} 1 \\ 0 \end{pmatrix}$

$2 + a = 5$.
Here a is an ordinary number, known sometimes as a ***scalar***, $a = 3$.

$\begin{pmatrix} 1 \\ 4 \end{pmatrix} + \begin{pmatrix} 2 \\ 3 \end{pmatrix} = \mathbf{a}$. Here **a** is a vector, $\mathbf{a} = \begin{pmatrix} 3 \\ 7 \end{pmatrix}$

Vectors as displacements

Vectors can be used to describe journeys or ***displacements***.

Example:

The journey shown in green is 3 across and 2 up, or as a vector, **d**

$\mathbf{d} = \begin{pmatrix} 3 \\ 2 \end{pmatrix}$

Vector quantities

Some physical quantities are described by both a size and a direction. These quantities are vector quantities.

Quantities like time, which have no direction, are scalar quantities.

Examples:
Displacement is a vector but distance and length are scalar quantities.
Velocity is a vector but ***speed*** is a scalar.

Vectors to describe translations

In a ***translation*** every point moves the same displacement, so ONE vector will describe the translation.

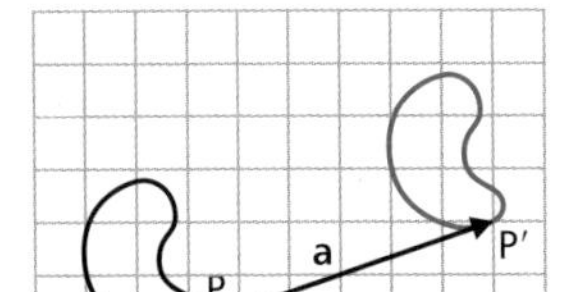

Example (left):
The black shape is given a translation. The ***image*** is shown in green. For any point P, the displacement P to P′ is described by the vector

$\mathbf{a} = \begin{pmatrix} 6 \\ 2 \end{pmatrix}$.

So this translation is the translation $\begin{pmatrix} 6 \\ 2 \end{pmatrix}$.

Scalar multiplying of a vector

When a vector is multiplied by a scalar (an ordinary number) each item in the vector is multiplied by the scalar.

$$k \times \begin{pmatrix} a_1 \\ a_2 \\ a_3 \end{pmatrix} = \begin{pmatrix} ka_1 \\ ka_2 \\ ka_3 \end{pmatrix}$$

Examples:

If $\mathbf{a} = \begin{pmatrix} 2 \\ 1 \\ 7 \end{pmatrix}$, then $3\mathbf{a} = 3\begin{pmatrix} 2 \\ 1 \\ 7 \end{pmatrix} = \begin{pmatrix} 6 \\ 3 \\ 21 \end{pmatrix}$

If $\mathbf{p} = k\mathbf{q}$, where k is a scalar,
then **p**, shown in green, is k times the length of **q** and ***parallel*** to **q**.

Vector addition

The addition of column vectors can be performed by adding corresponding items.
Example:

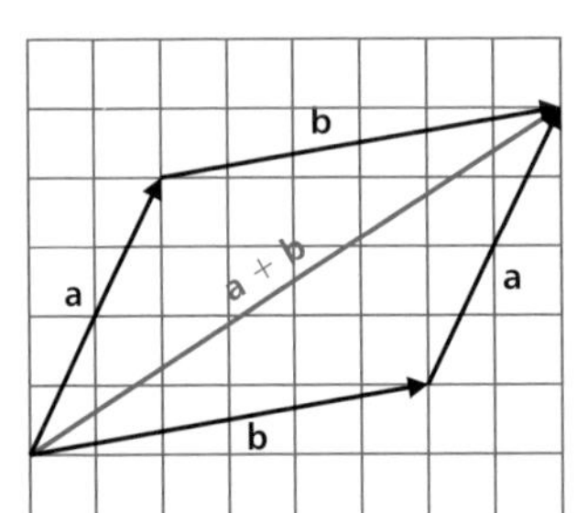

If $\mathbf{a} = \begin{pmatrix} 2 \\ 4 \end{pmatrix}$ and $\mathbf{b} = \begin{pmatrix} 6 \\ 1 \end{pmatrix}$ then $\mathbf{a} + \mathbf{b} = \begin{pmatrix} 2 \\ 4 \end{pmatrix} + \begin{pmatrix} 6 \\ 1 \end{pmatrix} = \begin{pmatrix} 8 \\ 5 \end{pmatrix}$

The addition of any two vectors can be done using the ***triangle law***.

NOTE: The addition of vectors is ***commutative***, $\mathbf{a} + \mathbf{b} = \mathbf{b} + \mathbf{a}$

Cartesian components of a vector

In the ***Cartesian plane***, **i** is a ***unit vector*** parallel to the *x-**axis*** and **j** is a unit vector parallel to the *y*-axis.

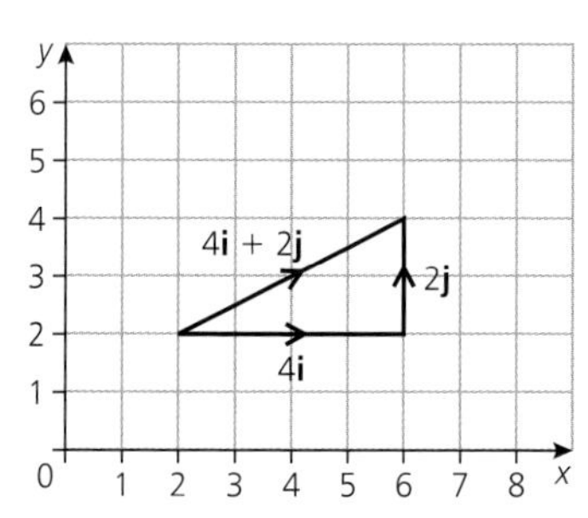

The vector $\mathbf{a} = \begin{pmatrix} 4 \\ 2 \end{pmatrix}$ can be written as $\mathbf{a} = 4\mathbf{i} + 2\mathbf{j}$.

4**i** and 2**j** are called the Cartesian components of **a**.

In three-dimensional space, **k** is a unit vector parallel to the *z*-axis.
$\mathbf{a} = 2\mathbf{i} + 4\mathbf{j} + 2\mathbf{k}$.
2**i**, 4**j** and 2**k** are the Cartesian components of **a**.

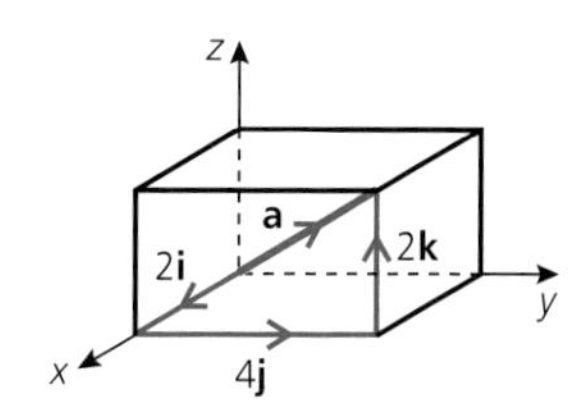

Velocity

Velocity is a measure of the ***speed*** with which a body moves in a particular direction. Velocity is a ***vector*** quantity.

Example: A car whose speed is 40 metre/second has a velocity of 40 metre/second due north if it is travelling in a northerly direction. If the car then proceeds in an easterly direction with the same speed, 40 metre/second, it has a different velocity.

Venn diagram

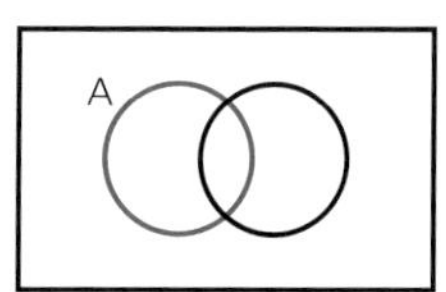

A Venn diagram is used to show ***sets***. Each set is represented by the ***region*** inside a simple closed curve, usually a ***circle***, and the ***universal set*** is shown by a ***rectangle***.

Example:
Any ***element*** of set A must be INSIDE the green circle.
Any element not in set A must be OUTSIDE the green circle.

Example:

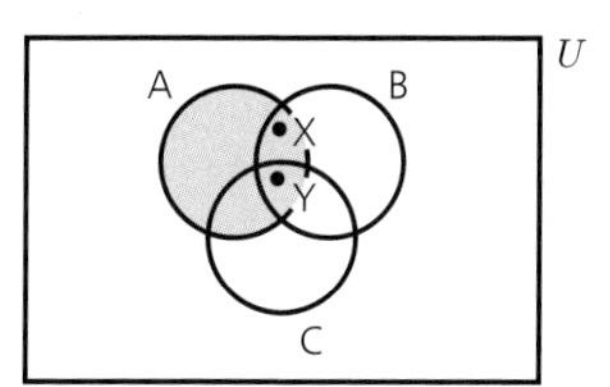

The set A is all the points in the ***region*** shaded green.
X and Y are in set A.
X is also in set B.
Y is also in both set B and set C.

Vertex (plural **Vertices**)

A vertex is a corner of a shape (or solid). In a ***polygon*** it is the point where the sides meet.

Examples:

The vertices of these shapes are shown by green dots.

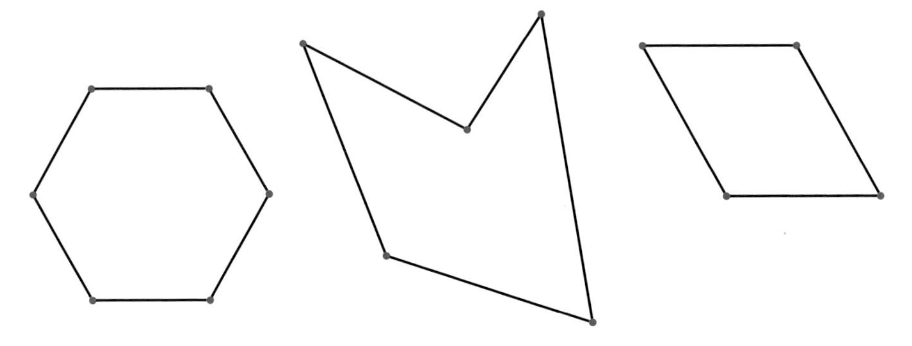

In a ***polyhedron***, it is the point where the ***edges*** meet.

Examples:
All the vertices of this ***cube*** are shown by green dots.

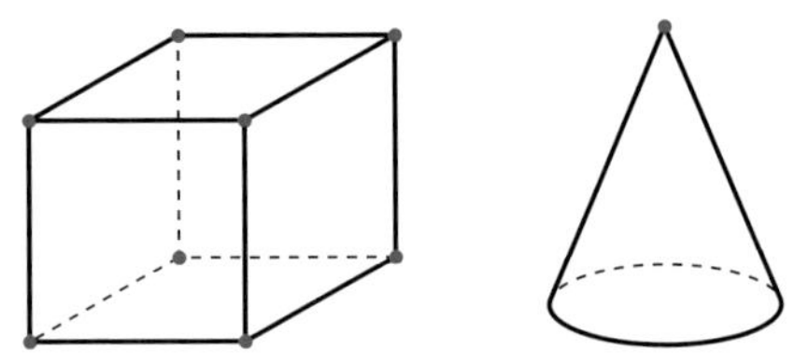

A ***cone*** only has one vertex, which is shown on the cone above by a green dot.

Vertical

A ***line*** is vertical if it is ***parallel*** to a string with a weight hanging on the end. Builders use a weight on the end of a string so that the walls can be constructed vertically.

Example (left): The green line shows the vertical line through the string.

NOTE: Any object which is dropped follows a line which is vertical.